A World Compendium

The GM Crop Manual

Editor: L. G. Copping

Promoting the Science and Practice of Sustainable Crop Production

British Library Cataloguing in Publication Data. A catalogue record of this book is available from the British Library.

ISBN 978-1-901396-19-5

Printed by Latimer Trend & Co., Plymouth

Cover photograph supplied by BeeWild Photography

Published by:
BCPC, 7 Omni Business Centre, Omega Park, Alton, Hampshire, GU34 2QD, UK
Tel +44 (0)1420 593 200 Fax +44 (0)1420 593 209
Email: md@bcpc.org Web: www.bcpc.org

All BCPC publications can be purchased from:
BCPC Publications, 7 Omni Business Centre, Omega Park, Alton, Hampshire, GU34 2QD, UK
Tel +44 (0) 1420 593 200 Fax +44 (0) 1420 593 209
Email: publications@bcpc.org
Or direct from the BCPC Online Book Shop at www.bcpc.org/shop

Disclaimer

Every effort has been made to ensure that all information in this edition of *The GM Crop Manual* is correct at the time of going to press. However, the editor and the publisher do not accept liability for any error or omission in the content, or for any loss, damage or any other accident arising from the use of the products listed therein.

The fact that a genetically modified crop has been approved in a country does not necessarily mean that it is grown, sold or processed there.

Contents

The Publisher

The GM Crop Manual is published by BCPC, a registered charity, formed in 1967.

The principal aim of BCPC is 'to promote and encourage the science and practice of crop production for the benefit of all'.

BCPC brings together a wide range of organisations interested in the improvement of crop production. The members of its Board represent the interests of government departments, the agrochemical industry, farmers' and consumers' organisations, the advisory services and independent consultants, distributors, the research councils, agricultural engineers, environment interests, training and overseas development.

The corporate members of BCPC currently are:

- Agricultural Engineers Association
- Association of Applied Biologists
- Association of Independent Crop Consultants
- Biotechnology and Biological Sciences Research Council
- British Institute of Agricultural Consultants
- British Society for Plant Pathology
- Campden BRI
- Crop Protection Association
- Department for Environment, Food and Rural Affairs, represented by the Chemicals Regulation Directorate (formerly Pesticides Safety Directorate)
- Department of Agriculture and Rural Development for Northern Ireland
- Environment Agency
- Imperial College, London
- Lantra
- National Association of Agricultural Contractors
- National Consumer Federation
- National Farmers' Union
- National Institute of Agricultural Botany
- Natural Environment Research Council
- Scottish Government Directorate General Environment
- Society of Chemical Industry – BioResources Group

Preface

We live in difficult times. It is predicted that the world's population will exceed 9 billion by 2050, and even today a great many people are starving and many more are malnourished. Some argue that the way forward is to teach the less developed nations how to cultivate their land and grow crops more effectively using traditional techniques. But the question of how this can be translated into greater food production is rarely addressed. In many less developed countries, the areas that are cultivated are determined by the amount of land a family can weed. If the family is debilitated by malaria and HIV AIDS, this area is very small indeed, and it is unlikely that the yield will be sufficient to feed the family. This means that the family are undernourished and the deadly circle of starvation continues. Is the answer to expect these poor people to work harder for less return, or is it to give them a better chance with the introduction of new technology?

Genetically modified (GM) crop technology, of course, will not only benefit the poorer nations; it will revolutionise agricultural practice throughout the world. Over 10 years of GM crop cultivation has shown significant financial advantages for the farmer and environmental advantages for the global population. In *The UK Pesticide Guide 2010*, for example, there are well over 40 herbicides and herbicide combinations that have approval for weed control in sugar beet, and many of these are recommended as multiple applications. If growers planted 'Roundup Ready' or 'Liberty Link' cultivars, they would need only a single herbicide to control all the potential weed problems, albeit with two applications to ensure total weed control. These GM crops allow the farmer to use a single (or very few) herbicides, and these are herbicides with very sound toxicity and environmental profiles. The benefits are lower input costs, better weed control, reduced residue levels in crops, and lowered environmental impact. What are the downsides? It has been argued that the use of such herbicide-tolerant varieties ransoms the grower to major agrochemical companies, who supply both the seed and the herbicides. This is an oversimplification, as most herbicides are no longer covered by patent, and generic versions are just as effective as branded products. In addition, the genes that allow the use of non-selective herbicides selectively are also in the public domain and most companies, large and small, have access to the technology. Is this so different from the sale of conventional crop protection agents?

Another often-used argument is that of the onset of resistance. For this reason, the US Department of Agriculture put in place a strategy of planting refugia to allow phytophagous insects access to crops that did not contain, and were not treated with, Bt or other insect-active proteins. This markedly reduced the possibility of the onset of resistance and it was a required strategy for all US farmers. Today's insect-resistant crops contain a number of different protein toxin genes with a view to both increasing the spectrum of insect control and reducing the possibility of resistance developing. Resistance to glyphosate in the light of the huge area planted to 'Roundup Ready' crops has been noticed, but is manageable with rotations and the use of either stacked herbicide traits or other crop-selective products.

And the future is even brighter with the possibility of introducing output traits that will give untold benefits to mankind. We already have the yellow rice from Syngenta (donated to the less developed world) containing genes that allow the crop to produce β-carotene, a precursor of vitamin A, the lack of which causes blindness in children. If we wish to feed the world in 2050, we will have to cultivate more crops or increase the yield of those we grow at present. Quite correctly, there are demonstrations against the felling of forests to give more land to agriculture, and this cannot be contemplated as an answer to the food shortage problem. We have been offered two possible solutions. The first is to concentrate all crop production on the most productive land with all the necessary inputs to guarantee maximum productivity. The less productive land can then be left for the benefit of biodiversity, with no inputs and no human-led interventions. The second is to consider land that is currently non-productive because it is too dry, too wet, too hot, too cold, too saline. There is much work ongoing in attempts to produce crops that can be grown in conditions that are usually unwelcoming and that always lead to reduced yields. None is currently commercialised, but this cannot be far away.

Two other areas of activity that fall outside the remit of this volume are the production of biofuels, and using transgenic plants as factories for pharmaceuticals and other fine chemicals.

Much of this work is being undertaken in countries such as China and India, and it is always difficult to know what is under development, how the research is being undertaken, and how close these crops are to market. It is hoped that these countries will become more open as the technology develops, and that many of these products can be included in the second edition.

This Manual outlines all the currently available GM crop products demonstrating both input and output traits. It is as up-to-date as a volume of this type can be at the time of going to press. However, the field is developing so rapidly that newer and better products will be launched to coincide with the new planting seasons. These new products will be included in the second edition.

Leonard G Copping
Editor

Introduction

Ever since humans first started keeping grains to use as seed in the next growing season, yields of crops have increased. Visual selection for desirable features unknowingly (at the time) selected for genes that would provide for further improvements. Thus, phenotypic selection following genetic crossing, combined with improved understanding of cultural and agronomic practices, allowed agriculture to feed an ever-increasing population. As the population grew into the billions, the pressure to produce more food with fewer resources intensified, as did primary crop production. Plant breeding became more complex and had elements of directed breeding, even although random-like recombination of genetic material was still not fully understood. It can be clearly shown that new genetics and new technologies have increased productivity per unit land area.

Using a global crop basket including cereals, sugar, citrus, fruits, oilcrops, pulses, roots & tubers and vegetables, FAO data were averaged for area harvested and average yield. The chart shows the increase in yield compared with the much lower increase in land used, over the past 15 years.

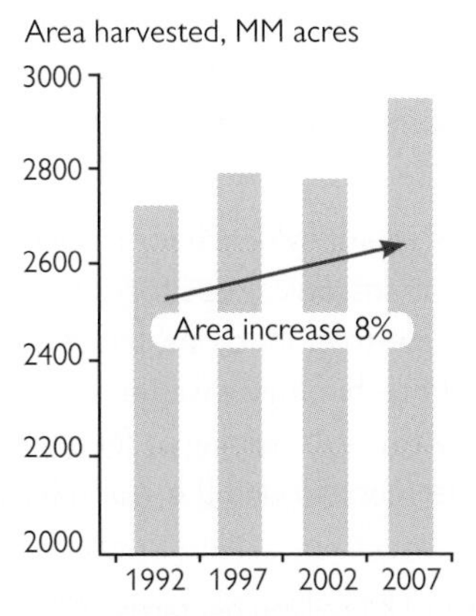

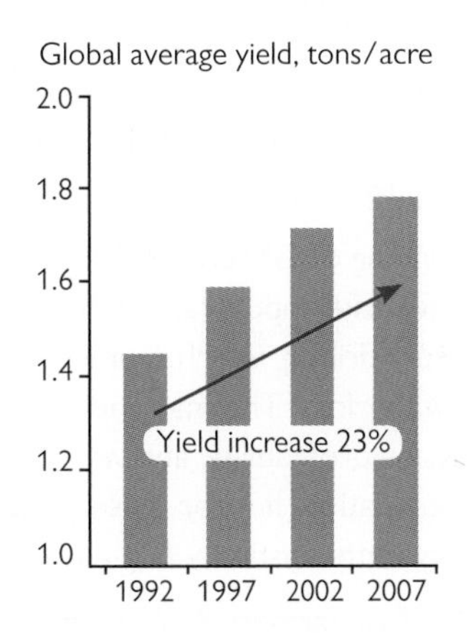

Source: Food and Agriculture Organization

In the past decade, the projections for population growth suggest there will be 9 billion people to feed by 2045 (*World Population Prospects: The 2008 Revision*, United Nations, Department of Economics and Social Affairs). Yet again, crop production stands ready to meet the increasing demand for food, and for healthier and more sustainable food supply. The technology is available in the form of agbiotechnology, which allows specific desirable traits to be developed via the directed manipulation of genes within a breeding programme.

The combination of advanced germplasm creations via the application of genomic markers with the transgenic introduction of key genes that impact the phenotypic expression of the desired trait has proved to be a critical aspect of modern crop production. Agbiotech has moved the yield curve in a positive manner, and we now have good prospects to be able to feed a growing population without having to dig up roads and shopping centres to remake fields, and with less impact on the environment.

Despite the tremendous success of agbiotech, a few small social groups have generated a crusade to inhibit the implementation of transgenic crops (also called genetically modified or GM crops) and associated new technologies for crop improvement. Without any real scientific evidence, anti-GM campaigners claim that the introduction of transgenic crops is bad because they are somehow 'unnatural', or that they may spread into the wider environment with an 'inevitable negative effect' on biodiversity, or that their safety has not been demonstrated, or that they are another way for big business to extort money from farmers, or that they will actually increase the cost of farming.

There are, of course, counter-arguments from supporters of GM technology, which are supported by tested scientific investigation rather than being based on belief. It has been shown that the farm-level economic effects, the production effects, the environmental impact resulting from changes in the use of insecticides and herbicides, and the contribution towards reducing greenhouse gas emissions of biotech trait adoption are all of benefit to the environment and humanity. This research is based on, and draws from, an extensive literature review of impacts globally, coupled with some collection and analysis of data, notably relating to pesticide usage.

Since the beginning of GM crop cultivations, there is sound evidence for a significant reduction in pesticide inputs (and hence residues) in transgenic crops together with yield increases. Direct drilling, which is practised in herbicide-tolerant crops, has reduced soil erosion and water loss. The many advantages of GM technology ease the workload of farmers, increase profitability, and will go a long way towards helping to feed the world's burgeoning population. It is for these reasons that farmers around the world have accepted the technology enthusiastically.

A record 14 million farmers, in 25 countries, planted 134 million hectares (330 million acres) of transgenic crops in 2009, a significant increase of 7% or 9 million ha (22 million acres) over 2008.

Global area of biotech crops in 2009 by country

Rank*	Country	Area (million ha)	Biotech crops
1	USA	64.0	Soybean, maize, cotton, canola, squash, papaya, alfalfa, sugar beet
2	Brazil	21.4	Soybean, maize, cotton
3	Argentina	21.3	Soybean, maize, cotton
4	India	8.4	Cotton
5	Canada	8.2	Canola, maize, soybean, sugar beet
6	China	3.7	Cotton, tomato, poplar, papaya, sweet pepper
7	Paraguay	2.2	Soybean
8	South Africa	2.1	Maize, soybean, cotton
9	Uruguay	0.8	Soybean, maize
10	Bolivia	0.8	Soybean
11	Philippines	0.5	Maize
12	Australia	0.2	Cotton, canola
13	Burkina Faso	0.1	Cotton
14	Spain	0.1	Maize
15	Mexico	0.1	Cotton, soybean
16	Chile	<0.1	Maize, soybean, canola
17	Colombia	<0.1	Cotton
18	Honduras	<0.1	Maize
19	Czech Republic	<0.1	Maize
20	Portugal	<0.1	Maize
21	Romania	<0.1	Maize
22	Poland	<0.1	Maize
23	Costa Rica	<0.1	Cotton, soybean
24	Egypt	<0.1	Maize
25	Slovakia	<0.1	Maize

*The first 15 are biotech mega-countries growing 50,000 ha, or more, of biotech crops.

Source: Food and Agriculture Organization

Clearly, farmers are applying GM technology around the world, with the exception of a few regions, notably Europe. But what genes are available to growers, and where are they cleared for use? This volume addresses these questions.

The regions that have approved and adopted transgenic crops have seen widespread benefits to farmers, and this is not just in the USA. Of the $51.9 billion additional gain in farmer income generated by biotech crops in the first 13 years of commercialisation (1996–2008), it is noteworthy that half, $26.1 billion, was generated in developing countries and the other half, $25.8 billion, in industrial countries.

China launched a $3.5 billion research project for GM crops to help address the demand for food in the world's most populous nation in 2008, after Premier Wen Jiabao told senior scientists that the country needs big science and technology measures like GM to solve its food problem. China has already widely planted insect-resistant GM cotton, which occupies 70% of the area devoted to growing the crop in China. Chinese scientists have also successfully developed several types of GM rice, whose field trials have shown higher yields and reduced pesticide use, although the government has delayed commercialisation of GM rice due to biosafety concerns.

China wants to push forward with the large-scale planting of GM crops, according to its first policy document of 2010, with Bt rice and phytase maize (which eliminates the need to feed extra phosphate to poultry and pigs) expected to be widely grown within 3–5 years. The 2010 'Number One Document', a publication issued annually by both the ruling Communist Party and State Council, which sets the agenda for that year's major work, was published on 31 January. It says that China will 'industrialise' GM crop farming. This is the seventh such document since 2004 to have concentrated on agricultural development.

The development of new GM crops is one of the 16 major projects listed in China's plan for scientific and technological development until 2020. The government's plans include the development of pest- and disease-resistant GM rice, rapeseed, maize and soybean, with research focusing on yield, quality, nutritional value and drought tolerance.

A great deal of work is under way in both China and India, but it is difficult to determine exactly what genes are being used, what crops transformed, what safety tests have been conducted, and what crops are being sold. For this reason, no crops from China or India are included in this edition of *The GM Crop Manual*, but it is expected that this information will be available for the second edition.

It is usual to refer to two basic types of GM crop – those showing input traits and those demonstrating output traits. Input traits are designed to modify the treatments or inputs into commercially grown crops, whilst output traits are designed to show a changed harvestable yield or modified outputs. In addition, there is a great deal of work under way around the world looking for ways of modifying crops so that they will grow in arid or salty regions, or show improved photosynthetic capability or reduced allergenic symptoms.

Input traits

Insect-resistance traits

Most current insect-resistance traits have been developed using *cry*, or crystal, genes from the bacterium *Bacillus thuringiensis* (Bt). The crystal proteins produced by the bacterium are highly toxic to a broad range of insect pests, but are not harmful to mammals or other organisms. Bt *cry* genes were a natural choice for an insect protection trait, as crystal protein and spore fermentation products have been used successfully as sprayable insecticides for many years. Transgenic varieties of cotton ('Bollgard 1', Monsanto) and corn ('YieldGard 1', Monsanto) transformed with the Bt genes have been on the market for nearly 10 years. These products are effective in controlling lepidopteran insects, and have significantly reduced the field use of insecticides. 'Herculex 1' (*cry1F*, *pat*; Dow AgroSciences and Pioneer Hi-Bred International) maize has broad-spectrum control that includes corn borers, black cutworm and other lepidopteran pests. The Chinese Academy of Agricultural Sciences has also developed and commercialised insect-resistant cotton varieties using the *cry1Ac* gene, alone or in combination with the cowpea trypsin inhibitor gene *CpTI*. Potato varieties resistant to Colorado beetle were developed by Monsanto using the Bt gene *cry3A*. After receiving regulatory approval in 1995, insect-protected 'NewLeaf' potato varieties were sold in the USA and Canada for several years. However, production of transgenic potatoes was discontinued in 2001 after a new effective insecticide (imidacloprid) was introduced and several major food processors announced that they would not use transgenic potato varieties. 'StarLink 1' (*cry9C*) maize, developed by AgrEvo (now Bayer CropScience) and approved only for feed use, was grown commercially in 1999 and 2000, but was discontinued after the gene was found in food products. 'Bt-Xtra 1' (*cry1Ac*; DeKalb Genetics Corp., DeKalb, IL [now Monsanto]) and 'NaturGard'/'KnockOut' (*cry1Ab*; Ciba Seeds [now Syngenta, Basel, Switzerland] in collaboration with Mycogen [now Dow AgroSciences]) maize varieties were also commercialised, but have since been replaced by newer products that offer increased insect protection throughout the growing season.

Second-generation products with stacked input traits started entering the market in 2003 with the introduction of 'Bollgard II' (*cry1Ac*, *cry2Ab2*) cotton by Monsanto. Features of second-generation products include stacked genes (e.g. combined traits of herbicide and insect resistance), two modes of action (e.g. two different Bt genes combined in one product) for improved insect resistance management, and enhanced performance of the traits (e.g. increased spectrum of target insects). For example, 'WideStrike' (*cry1Fa*, *cry1Ac*; Dow AgroSciences) cotton and 'YieldGard Plus' (*cry1Ab*, *cry3Bb1*; Monsanto) maize are newly released products with stacked Bt gene traits. Triple-stack traits – dual Bt genes combined with glyphosate or glufosinate resistance – are likely to replace many of the single-trait products.

Syngenta International AG announced on 13 November 2009 that its genetically modified maize traits MIR162 and Bt11 × GA21 are now fully approved for cultivation in Brazil; on 3 December 2009 that it had received cultivation approval for its maize trait GA21 in the Philippines; and on 22 December 2009 that it had received cultivation approval for its genetically modified maize trait Bt11 × GA21 from the Ministry of Agriculture in Argentina. The efficacy and broad spectrum of MIR162 make it an advanced technology available to control the fall armyworm (*Spodoptera frugiperda* J.E. Smith). The stacked maize trait Bt11 × GA21 offers higher value and convenience for growers through combined insect resistance and herbicide tolerance. The GA21 trait genetically enhances maize seeds with a built-in resistance to herbicides, offering growers increased productivity and reduced production costs.

Dow AgroSciences has gained registration for 'WideStrike' Insect Protection in Brazil and it has advanced 'WideStrike' to Phase 2 regulatory trials in India. Dow AgroSciences plans to introduce its new herbicide-tolerant trait technology in the USA in cotton in 2015.

Lepidopteran-resistant 'VipCot' cotton products, with the unique Vip3A vegetative insecticidal protein from Bt, will also be introduced soon by Syngenta in collaboration with Delta and Pine Land Company. Syngenta also has a corn rootworm product (modified *cry3A*) with regulatory approval pending. Internationally, anticipated new transgenic products include insect-resistant cotton hybrids in India and insect-resistant rice in China and Iran, all developed using Bt genes.

Herbicide-tolerant crops

Among the input traits offered to farmers, herbicide resistance has been the most widely adopted. Glyphosate-tolerant soybeans were one of the earliest transgenic crops brought to market and they have experienced rapid adoption, to the point that over 85% of US soybeans and 56% of soybeans globally are now glyphosate tolerant. The use of glyphosate-tolerant cotton, canola and maize is also increasing at a rapid pace, especially when combined, or stacked, with insect-resistance traits. Glyphosate tolerance is achieved in 'Roundup Ready' brands (Monsanto) by expression of a modified *Agrobacterium* gene coding for the herbicide-insensitive enzyme CP4 enolpyruvyl-shikimate-3-phosphate synthase (CP4 EPSPS). The NK603 trait has two copies of *cp4 epsps* with different promoters for better expression in the meristems. Glyphosate herbicides are relatively inexpensive and can be applied over the top of tolerant seedling crops. Nearly all broadleaf and grass weeds are eliminated, resulting in reduced competition, higher yields and cleaner fields at harvest. Adoption of reduced and no-till practices, where dead vegetation is left in the field rather than ploughed under, has been a significant unintended feature of herbicide-tolerant crops, saving farmers money in fuel costs and reducing soil erosion. 'Roundup Ready Flex' (Monsanto) was deregulated by the US Department of Agriculture (USDA) in 2005. 'Flex' cotton contains a second expression cassette of *cp4 epsps* with a meristem-active promoter, which allows an extended application window and higher glyphosate application rates. 'Roundup Ready Wheat' was deregulated and approved for sale, yet has not been sold owing to lack of support by the wheat industry. It will probably be sold when additional traits for disease resistance or nutritional quality are also available. One 'Roundup Ready Sugar Beet' trait was approved for sale in the USA, but faced opposition from sugar refineries and food manufacturers. Another 'Roundup Ready' sugar beet has been deregulated in the USA and, pending approval in major growth and export countries, was first marketed in 2007. Glyphosate tolerance has become a must-have trait for major seed companies.

Recently, Origin Biotechnology has reached a comprehensive, worldwide agreement with the Institute of Microbiology of the Chinese Academy of Sciences and Sichuan Biotech Engineering Ltd, who jointly own the rights to an internally developed gene that gives high tolerance to glyphosate. This glyphosate-tolerance gene, demonstrated to be extremely effective in both laboratory and field environments, is entirely new to consumer markets in that it has never been commercialised, and is protected by patents granted separately by China and the USA. Origin Biotechnology has exclusive rights to sell and develop maize, soybean, rice, cotton and canola products that contain these technology traits worldwide. It is expected that these traits will appear in the second edition of this Manual.

On 22 May 2009, the USDA granted approval for the 'GlyTol' cotton technology developed by Bayer CropScience. Cotton varieties with the 'GlyTol' trait are tolerant to the herbicide glyphosate. The USDA's decision, which follows approval from the US Food and Drug Administration, marks another milestone towards the first commercialisation of 'GlyTol' cotton. The 'GlyTol' trait provides robust tolerance to applications of glyphosate, and gives growers the flexibility to select any brand of glyphosate herbicide labelled for use on cotton without concern for crop safety.

The first alternative mode of action trait for glyphosate tolerance, detoxification by glyphosate *N*-acetyltransferase, is being developed by Pioneer Hi-Bred and is expected to be offered in combination with tolerance to sulfonylurea herbicides. Dual herbicide-tolerance products will offer all the advantages of glyphosate tolerance with the additional benefits of two modes of action for resistant weed management.

Traits for tolerance to three other classes of herbicide have been developed, but have not reached the same level of popularity as glyphosate tolerance. Tolerance to oxynil herbicides conferred by the BXN nitrilase from *Klebsiella pneumoniae* (Abel) Bergey *et al.* (subspecies *ozaenae*) was the first trait engineered in cotton (developed by Calgene [now Monsanto]). Because glyphosate is less expensive and controls more weed species, interest in using the oxynil herbicides has waned, and 2004 was the final year of BXN cotton sales. BXN canola was commercialised by Rhone-Poulenc Canada (now Bayer CropScience) and then discontinued. Phosphinothricin acetyltransferase (PAT or BAR) detoxifies phosphinothricin- or bilanafos-based herbicides (glufosinate). The *pat* gene is native to *Streptomyces viridichromogenes* and *bar* is from *S. hygroscopicus*, where they act in both the biosynthesis and detoxification of the tripeptide bilanafos. Like glyphosate, phosphinothricin herbicides control a broad spectrum of weed species and break down rapidly in the soil so that problems with residual activity and environmental impact are greatly reduced. Bayer CropScience markets this trait as 'LibertyLink' in several crops. The *pat* and *bar* genes are also popular plant transformation markers in the research community. Pending approval in the European Union, 'LibertyLink' soybeans (Bayer CropScience), first approved in the USA in 1998, might soon be sold commercially. Finally, BASF markets non-transgenic 'CLEARFIELD' imidazolinone-resistant canola, wheat, sunflower, maize, lentil and rice, while DuPont markets non-transgenic STS soybean with tolerance to sulfonylurea herbicides. A sulfonylurea-tolerant flax variety called 'CDC Triffid', developed by the University of Saskatchewan, was grown commercially in Canada in 2000, but is no longer offered. These crops all contain modified versions of the acetohydroxy acid synthase gene, also called acetolactate synthase (ALS), which is not inhibited by imidazolinone and/or sulfonylurea herbicides. Herbicides that inhibit ALS are considered low or very low use-rate herbicides with a good spectrum of weed control, and are likely to remain an important part of weed resistance-management programmes.

2,4-D mono-oxygenase is being used to transform elite Australian cotton varieties that are being developed through traditional breeding techniques.

Other herbicide traits with new modes of action are currently in the research phase. Recent publications describe herbicide resistance using glutathione conjugation, *p*-hydroxyphenylpyruvate dioxygenase-inhibiting herbicide resistance, protoporphyrinogen oxidase-inhibiting herbicide resistance, and broad-range herbicide tolerance owing to P_{450}-catalysed detoxification. Detoxification of dicamba herbicides could translate to herbicide-tolerant plants as well.

Mixed traits

Several insect-protected products are nearly commercial or are at the final stage of development. For example, 'Herculex RW' (*cry34Ab1*, *cry35Ab1*, *pat*) and the broad-spectrum, triple-stack 'Herculex XTRA' (*cry1F*, *cry34Ab1*, *cry35Ab1*, *pat*) maize hybrid lines tolerant to glufosinate herbicides, lepidopteran pests and corn rootworm, co-developed by Dow AgroSciences and Pioneer Hi-Bred, were available for commercial sale in 2006. A second-generation 'Roundup Ready YieldGard RW' maize product (*CP4 epsps*, *cry3Bb1*; Monsanto) tolerant to the corn rootworm, but lacking the *nptII* antibiotic gene, has just been deregulated by USDA.

Bio.org predicts that input traits headed for the market in the next 6 years include 'Roundup Ready' lettuce and strawberries, 'LibertyLink' rice, insect-protected soybeans, apples and additional maize products.

The new 'Smartstax' hybrids were launched in 2010 and are expected to grow rapidly (up to 12–15 M ha) over the next 2–3 years. These hybrids have eight stacked genes with multiple modes of action. Due to dual modes of insecticidal action, the US Environmental Protection Agency has officially agreed that, in this situation, the size of the refugia can be decreased to 5% of the planted area.

Nematode-resistant crops

Protein TcdA isolated from *Photorhabdus luminescens*, a symbiotic bacterium associated with soil nematodes, is another example of a non-Bt insecticidal protein shown to confer insect resistance. With the impending ban on methyl bromide soil fumigation owing to environmental concerns, nematode control will be an emerging market that has not been previously addressed by transgenes.

Scientists at the Agricultural Research Service have created PA-559, the first root-knot nematode-resistant red-fruited habanero-type pepper. The pepper is also resistant to the peanut root-knot nematode and the tropical root-knot nematode. Field plantings conducted in Charleston over 2 years confirmed the pepper's pest resistance and showed that the fruit characteristics of PA-559 are comparable with those of currently available red-fruited habanero-type cultivars. PA-559 is a relative of another root-knot nematode-resistant cultivar, TigerPaw-NR, which was released by ARS in 2006. Both parental lines used to develop PA-559 are sister lines of TigerPaw-NR. PA-559 contains a dominant gene that gives the plant its pest-resistance trait. This makes the variety ideal for use as a parental line in breeding resistant cultivars, because breeders can be sure the plant's offspring will contain resistance. Although recommended for use by breeders as a parental line, PA-559 can also be used in commercial production without further development.

These traits are not included in this volume as they continue in development, but it is expected that they will be included in the second edition.

Virus-resistant crops

Monsanto developed varieties of potato that are resistant to potato leafroll virus (using the coat protein gene) and marketed them as 'NewLeaf Plus' through NatureMark, who also sell 'NewLeaf Y' for control of potato virus Y (again by using the coat protein gene).

'SunUp 63-1' and 'SunUp 55-1' were developed in Hawaii for control of papaya ringspot potyvirus with the use of the coat protein gene. Asgrow Seed has released 'Freedom II' Squash for control of cucumber mosaic virus cucumovirus (CMV), zucchini yellows mosaic potyvirus (ZYMV) and water melon mosaic virus 2 potyvirus (WMV2) with the use of squash virus coat protein genes.

Genetically modified virus-resistant potato and peanut are under various stages of trials in the glasshouse and confined fields of select universities and research institutions in India. The programme is spearheaded by the Agricultural Biotechnology Support Project-II (ABSP-II) of Cornell University.

Newly emerging RNAi technology may provide additional traits such as virus resistance with broad-spectrum control and potentially enhanced durability.

Disease-resistant crops

By introducing a synthetic gene that codes for maiginin 2, researchers at the New Zealand Institute for Plant & Food Research Ltd developed potato plants resistant to *Erwinia carotovora*. The soil-dwelling microbe causes soft-rot disease in potatoes, carrots and other vegetables, and infections often result in a complete crop loss. The soft-rot-resistant potato plants that the New Zealand scientists developed express a synthetic *maiginin 2* gene. First identified in frog skin, maiginin peptides are selectively toxic to microbes and not mammalian cells. Several studies have also shown that the peptide has a broad activity against numerous phytopathogens, including some fungi and the bacterial agents that cause common scab and blackleg. In engineering the *maiginin* gene, the researchers made several mutations to reduce the peptide's susceptibility to proteolytic cleavage and increase its activity against prokaryotes. The transgenic potato lines were tested for three planting seasons. The soft-rot-resistant potatoes were found to be similar to conventional potato varieties in terms of yield and other agronomic performance criteria. It is expected that disease-resistant bananas and canola will be commercially available soon.

Syngenta is looking for plant-derived proteins with antifungal activity, with the intention of isolating the genes and using them to transform crops.

Significant progress has been made in the past decade in understanding the molecular mechanisms of plant resistance to diseases. From the picture of extremely complex plant–pathogen relationships, several potential approaches for developing disease control traits are emerging. The recent example of combining marker-assisted selection with genetic transformation to develop blast and bacterial blight resistance in rice illustrates how transgenic traits can be stacked with native resistance genes for broad-spectrum disease resistance.

Output traits

It may be argued that it is output traits that will revolutionise GM crop technology. Still in their early stages, there are many opportunities that are close to commercialisation. Typical approaches are listed below.

Phytase maize

Origin Biotechnology is working on phytase maize, which remains the only biotechnology maize product in Phase V of development in China. The worldwide phytase potential market size is $500m, including $200m for China alone, according to the China Feed

Industry Study. This will be the world's first transgenic phytase maize, and is expected to be commercialised as the first genetically modified maize product in China. Phase V passage is expected shortly pending a final-stage approval from the Ministry of Agriculture. Phytase is currently used as an additive in animal feed to break down phytic acid in maize, which holds 60% of the phosphorus in maize.

Drought resistance

Genetically modified drought- and salinity-tolerant rice are under various stages of trials in glasshouse and confined fields of select universities and research institutions in India. The programme is spearheaded by the Agricultural Biotechnology Support Project-II (ABSP-II) of Cornell University, which helped the development of the fruit and shoot borer-resistant brinjal and led it to the Genetic Engineering Approval Committee gateway for commercialisation. Professor Ray Wu of Cornell University has demonstrated that stress tolerance in plants can be induced by manipulating the genes that are responsible for the accumulation of the sugar 'trehalose'. Professor Wu's system is designed in such a way that the bioengineered genes are specifically turned on when the plant is under drought or salt stress. Through ABSP-II, the trehalose genes have been transferred to the Directorate of Rice Research, Hyderabad, to be introduced into local rice varieties and to evaluate the positive events in glasshouse conditions and screen them for drought tolerance. It is estimated that in India, 30% of the agricultural area receives less than 750 mm rainfall and is chronically drought-prone, and 35% of the area with 750–1125 mm rainfall is also subject to drought once in 4–5 years. ABSP-II estimates show that 68% of the total sown area covering about 142 million ha is vulnerable to drought conditions. Moreover, India accounts for nearly 47% of saline, 20% of sodic and 7% of acid sulfate soils of tropical Asia.

In another research project, a Chinese patent has been granted to FuturaGene, on the 'Method for increasing stress tolerance in plants'. The patent includes FuturaGene's drought-tolerance gene for use in both food and non-food crops, including the drought-tolerance technology granted by the company to Bayer CropScience for its utilisation in cotton worldwide. Since September 2009, the company has been collaborating with the Chinese Academy of Forestry on the use of the same gene technology for the development of new, enhanced poplar with increased water efficiency. China is the largest global cotton producer, by both volume and value, and is also a major potential market for drought-tolerant poplar, which could play an important role in reversing desertification. In a country with real concerns about food security, it is a major development in the process of establishing sustainable agriculture.

A team of scientists from Canada, Spain and the USA has identified a key gene that allows plants to defend themselves against environmental stresses such as drought, freezing and heat. The research team, led by Sean Cutler of the University of California, Riverside, has identified the receptor of the key hormone in stress protection, abscisic acid (ABA). Under stress, plants increase their ABA levels, which help them survive a drought through a process that is not fully understood. The area of ABA receptors has been a highly controversial topic in the field of plant biology that has involved retractions of scientific papers as well as the publication of papers of questionable significance.

An international team of researchers at the Australian National University has identified a gene in *Arabidopsis* that allows plants to survive drought. The gene, *sal1*, was found when they were looking at different mutant varieties of *Arabidopsis* that had unusual responses to high light. Mutations in *sal1* enable plants to survive longer without added water. The researchers are now in the process of introducing the mutant characteristics into the elite wheat cultivars currently used in the agriculture industry. The ultimate aim of the project is to develop wheat lines with improved drought tolerance and water use, and the next step will be to identify wheat mutant plants lacking *sal1* genes identified by molecular biology procedures. It is expected that the mutants should remain green, turgid and photosynthetically active, producing more leaves, flowers and seeds during mild to moderate water deficit. Because the basis of the mutation is a missing gene, it should be possible to create drought-tolerant wheat plants without resorting to transgenic methods. Drought-tolerant wheat plants may prove to be important in the future, as climate models predict that the vast wheat-growing areas of southern Australia will become drastically drier over the next 50 years.

More efficient photosynthesis

BASF PlantScience has signed a partnership licence and R&D agreement with the University of Koln's Botanical Institute covering plant biotechnology. Work will concentrate on plant traits that boost crop yields through better use of carbon dioxide and enhance crop tolerance of harsh conditions such as salinisation or drought.

Improved nitrogen utilisation

Arcadia Biosciences and Vilmorin have reached an agreement for the development and commercialisation of nitrogen use-efficient (NUE) wheat. Under the terms of the agreement, Vilmorin receives privileged global rights to the use of Arcadia's NUE technology in wheat. The combination of Arcadia's NUE technology and Vilmorin's genetic resources enables the development of high-yielding wheat that could require about half the amount of nitrogen fertiliser as conventional crops, offering economic benefits to growers and a measurable positive impact on the environment.

Modified starch and oil crops

On 11 June 2009, the European Food Safety Authority (EFSA) published its scientific assessment of a gene that is present in Amflora, as well as in other genetically modified products. In the past, EFSA has repeatedly assessed the use of this gene (*npt2*) and has again concluded that the gene is safe for humans, animals and the environment, and that no further scientific work is needed. The mandate to carry out an additional safety assessment on the *npt2* gene was given by the EU Commission in May 2008. The idea of developing 'Amflora' originated in the European starch industry. The aim was to improve industrial starch potatoes. Potatoes produce two types of starch: amylose and amylopectin. For many applications, such as in the paper, textile and adhesives industries, amylopectin is preferred. With plant biotechnology, BASF found a solution to avoid production of the unwanted starch component amylose. With its pure amylopectin starch, higher-quality starch can be extracted from 'Amflora'. The industry benefits as paper produced with high-quality 'Amflora' starch has a higher gloss, and concrete and adhesives can be processed for a longer period. Industrial processes can thus be optimised, which results in savings in raw materials such as water, additives and energy. Europe's farmers and the starch industry have estimated that the product and processing benefits, as well as the savings in raw materials and energy use, can be translated into a potential added value of €100–200 m per year. In addition, BASF expects peak income from licences in the range of €20–30 m per year.

Maize lines with unusual starch quality, particularly low onset temperature of gelatinisation, were developed by traditional plant breeding methods. Starting materials included exotic maize germplasm, adapted maize germplasm, and crosses between exotic and adapted germplasm. One year of yield testing of experimental hybrids of these lines showed acceptable agronomic performance. *Tripsacum dactyloides* is a wild relative of maize. Genes from *Tripsacum* can be introgressed into maize by cross-pollination and traditional plant breeding techniques. *Tripsacum* × maize populations were made to broaden the genetic base of Corn Belt hybrids. Lines from the populations were screened for grain quality traits including fatty acid composition of the oil. The plants recovered from the population with interesting fatty acid profiles were self-pollinated to develop lines. The plants were also crossed to both Stiff Stalk and non-Stiff Stalk Corn Belt inbreds. Fatty acid composition was evaluated at each stage of development and superior lines advanced, and backcrosses were made to Corn Belt inbreds. The fatty acids were converted to the methyl ester form and measured with gas chromatography. Fatty acid profiles were significantly enhanced in the *Tripsacum* introgressed ears compared with the Corn Belt inbreds. Oleic acid content was increased from 22.9% to 70.1%. Total saturated fatty acid content was decreased from 12.0% to 6.5% for low-sat lines. Total saturated fatty acids were increased from 12.0% to 24.0% for high-sat lines. Ear size was also significantly increased by crossing the recovered introgressed lines to Corn Belt inbreds and selecting for superior quality.

Allergy prevention

Japanese scientists have achieved a breakthrough in advancing towards the next generation of genetically modified rice, which will fight allergies instead of causing them. According to a report in *Live Science*, the new transgenic rice designed to fight a common pollen allergy appears safe in animal studies. Fumio Takaiwa and colleagues note that the first generation of GM crops was designed to help keep crops free of weeds and insects. The next generation of transgenic crops is being developed to benefit human health directly. The rice plant has been genetically engineered to fight allergies to Japanese cedar pollen, a growing public health problem in Japan that affects about 20% of the population. In laboratory studies, the researchers fed a steamed version of the transgenic rice and a non-transgenic version to a group of monkeys every day for 26 weeks. At the end of the study period, the test animals did not show any health problems, in an initial demonstration that the allergy-fighting rice may be safe for consumption, according to the researchers. More research will be needed to bring the rice to market.

Biopharming and nutraceuticals are not covered.

Conclusions

The first herbicide- and insect-resistance input traits are heading into their second decade of adoption. The acreage planted with input traits continues to increase for maize, soybean, canola and cotton in the USA and globally. Relatively few genes have been approved and the major crops dominate due to the high costs of bringing the traits to market. The current herbicide-tolerance, insect- and virus-resistance traits meet the major needs of farmers in developed nations. New diversity in modes of action towards the major target pests and herbicides will help prolong the usefulness of these traits. The scientific capabilities to develop useful crop traits and the desire of growers to use them are high. Nevertheless, several outstanding questions remain that could affect the future use of input traits. How will future products fare against public opinion and the challenges of the current regulatory systems? Will there be regional or global markets for these products that can justify their development expenses? Will acceptance of consumer-oriented traits facilitate acceptance of future input traits in Europe? Grower demands for transgenic traits may be high, but in global agriculture the (seed-buying) customer must also deal with the input from governments and society in determining the availability of crop traits. Future expansion in this area will depend strongly on the answers to some of these issues.

This Manual clarifies the traits available and indicates the safety and yield enhancement that GM technology brings. It will be of interest to consultants, growers, researchers, industry and academia. It will also be a major reference work for all those interested in the application of molecular biology to crop protection and crop production.

1 INPUT TRAITS

1.1 Disease resistance

1.1.1 *papaya ringspot virus* gene

Introduces resistance to papaya ringspot virus

NOMENCLATURE: Approved name: *papaya ringspot virus* gene.

Other names: *PRSV* gene.

Development code(s): 55-1 and 63-1; UFL-X17CP-6 (X17-2).

Promoter(s): 55-1 and 63-2 viral coat protein (*papaya ringspot potyvirus* (*PRSV*)) gene – CaMV35S; X17-2 viral coat protein (*papaya ringspot potyvirus* (*PRSV*)) – CaMV 35S *uidA* leader sequence.

Trait introduction: The transgenic papaya lines 55-1 and 63-1 were produced by biolistic (particle bombardment) transformation of embryogenic cultures of the papaya cultivar 'Sunset' with DNA-coated tungsten particles.

SOURCE: PRSV belongs to the potyvirus group and is an aphid-transmissible RNA virus that commonly infects papaya, causing serious disease and economic loss.

TARGET PESTS: Papaya ringspot virus.

TARGET CROPS: Papaya.

BIOLOGICAL ACTIVITY: In general, resistance conferred by the coat protein (CP) gene of a potyvirus (such as PRSV) is RNA-mediated and sequence-specific. Resistance is only effective when the transgene has a high similarity to the CP gene of the infecting virus and appears to be due to post-transcriptional gene silencing (PTGS). PTGS can take place when the antisense RNA, either directly from transcription of an antisense CP transgene or indirectly from sense transgene messenger RNA (mRNA) via RNA-dependent RNA polymerases, binds to the complementary regions of viral RNA transcripts in the cytoplasm to form duplexes, which are subsequently cleaved by dsRNA-specific nucleases into RNAs that interfere with transcription of homologous RNAs. This requirement for sequence homology points to a need to utilise local PRSV isolates as sources of the CP gene in order to obtain effective PRSV resistance in transgenic papaya.

COMMERCIALISATION: Formulation: Work is also underway at South China Agricultural University on developing PRSV resistant papaya, but no information is available on these varieties at the time of going to press (October 2010). **Trade names:** 'SunUp' and 'Rainbow' (University of Hawaii).

PRODUCT SPECIFICATIONS: The transgenic papaya lines 55-1 and 63-1 were field tested in the USA (1991–96). Extensive testing of papaya lines 55-1 and 63-1 in laboratory, glasshouse and field experiments determined that the plants exhibited the typical agronomic characteristics of the parent papaya variety 'Sunset', with the addition of resistance to

PRSV infection. The variation in agronomic characteristics did not differ significantly from that seen in commercial cultivars of papaya. Field trial reports demonstrated that the transformed papaya lines 55-1 and 63-1 had no effect on non-target organisms or the general environment.

MAMMALIAN TOXICITY: There is a history of safe consumption of PRSV Coat Protein (CP) derived from virus-infected papaya fruit, and there is no indication that the form of the CP expressed in transgenic papaya is materially different in any way that would affect its potential for toxic or allergenic effect. Additionally, the PRSV CP does not possess any of the physicochemical properties normally associated with protein allergens or toxins, such as heat stability and resistance to digestion by simulated gastric fluids. Also, there were no regions of homology when the deduced amino acid sequence of PRSV CP was compared with the amino acid sequences of known protein allergens or toxins.

ENVIRONMENTAL IMPACT AND NON-TARGET TOXICITY: It has been concluded that the genes inserted into transgenic papaya lines would not result in any deleterious effects or significant impacts on non-target organisms, including threatened and endangered species or beneficial organisms. The PRSV coat protein expressed in these papaya lines is found in all PRSV-infected plants, and there are no reports of this protein having any toxic effects. In fact, this viral coat protein is routinely ingested by virtually all animals, including humans, when papaya is consumed.

DATE(S) OF REGISTRATION: Environmental regulatory approval for 55-1 and 63-1 was granted in the USA in 1996; food and/or feed uses were approved in 1997; food use approval was given in Canada in 2003. Environment regulatory approval for X17-2 was granted in the USA in 2009; food and/or feed approval was given in 2008.

1.1.2 *Phytophthora infestans* resistance genes

Introduces resistance to Phytophthora infestans

NOMENCLATURE: Approved name: None – in development.

Other names: *Phytophthora resistance* genes.

SOURCE: A field trial of a genetically modified (GM) variety of potato resistant to late blight (*Phytophthora infestans* De Bary) began in eastern England in 2010. The trial, carried out by scientists from The Sainsbury Laboratory (TSL), will last 3 years. About 100 different species of *Solanum*, the genus to which potatoes belong, were tested and a few that were resistant to late blight were identified. Genes were isolated from two different wild potato species that confer blight resistance. Two of these genes, taken from inedible wild plants that grow in South America, were used to produce a genetically modified 'Desiree' variety. The resistance genes allow the plant to discern a pathogen attack as a cue to activate the host's defences.

1.1.3 *plum pox virus coat protein* gene

Introduces resistance to plum pox virus

NOMENCLATURE: Approved name: *Plum pox virus coat protein* gene.

Other names: *PPV CP* gene.

Development code(s): ARS-PLMC5-6 (C5).

Promoter(s): CaMV 35S.

Trait introduction: The coat protein gene of the plum pox virus containing the 35S promoter was subcloned into *Hin*dIII-digested pGA482GG and the resulting plasmid was designated pGA482GG/PPV-CP-33. This plasmid was used to electrotransform *Agrobacterium tumefaciens* strain C58/Z707 and used for transformation of plum tissue. A number of transgenic lines were produced by transformation of the plum cultivar 'Bluebyrd'.

TARGET PESTS: Plum pox.

TARGET CROPS: Stone fruit, especially plums.

BIOLOGICAL ACTIVITY: Mode of action: There is no detectable protein produced by the PPV coat protein gene and assays for the messenger RNA (mRNA) from this gene were also negative, prompting the deduction that the resistance mechanism is through post-transcriptional gene silencing (PTGS). This was confirmed by nuclear run-on analysis which confirmed that mRNA from the PPV coat protein was being produced, but was not accumulating in the plant cells. Another characteristic of PTGS is the production of short interfering RNA molecules (siRNA) that are derived from double-stranded RNA degraded by a specific RNAseIII-like enzyme referred to as DICER. It is these siRNA molecules which are thought to provide the specificity of PTGS in silencing certain genes. Detection of siRNA specific to the PPV coat protein genes confirmed this mechanism of transgenic virus resistance in C5.

COMMERCIALISATION: The US Department of Agriculture – Agricultural Research Service hopes to commercialise the C5 plum in the USA in the near future.

PRODUCT SPECIFICATIONS: Field trials were established in 1996–97 in Poland, Spain and Romania with C5 shoots grafted onto non-transgenic rootstocks. These trials were maintained for efficacy testing of the transgenic lines and deliberately inoculated with PPV using infected grafts or interspersed with infected trees to test for resistance to aphid-transmitted virus. Further field trials in the USA were approved in 1995 and included trees which were rooted from C5 clones as well as grafted material.

MAMMALIAN TOXICITY: There are no records of allergic or other adverse toxicological effects from researchers, breeders or growers of the transgenic varieties. The small-interfering RNAs (siRNA) responsible for the post-transcriptional gene silencing (PTGS) resistance mechanism in C5 plum are also not of concern as nucleic acids are a normal part of every living organism and do not have toxic or allergenic properties.

ENVIRONMENTAL IMPACT AND NON-TARGET TOXICITY: The biology of the C5 plum trees with respect to their potential to affect non-target organisms such as beneficial insects (including pollinators such as bees) and biocontrol organisms was evaluated. The C5 plum does not express detectable coat protein from PPV, which eliminates the concern of protein exposure to non-target organisms. Furthermore, even if C5 did express viral coat protein this would not increase the issue of potential impacts to non-target organisms as the PPV coat protein is not known to have any toxic properties. Plant viruses are ubiquitous in the environment and cause damage to fruits, leaves, seeds, flowers, stems, and roots of many important crop species. These viruses infect virtually every plant species and, under natural conditions, certain plant viruses are nearly always present on particular crop or weed hosts. Viral coat proteins are therefore routinely ingested by virtually all mammals when virus-infected fruits and vegetables are consumed.

DATE(S) OF REGISTRATION: In the USA, environmental regulatory approvals were obtained in 2007; and food and/or feed uses were approved in 2009.

1.1.4 *potato leafroll virus coat protein* gene

Introduces resistance to potato leafroll virus luteovirus

NOMENCLATURE: Approved name: *potato leafroll virus luteovirus coat protein* gene.

Other names: *PLRV coat protein* gene; *orf1 /orf2* gene; *orf1* (*replicase*); *orf2* (*helicase*) genes from PLRV

Promoter(s): Hel (helicase (*potato leafroll luteovirus* (*PLRV*) *orf* 2)) – figwort mosaic virus (FMV) 35S 77-nucleotide leader sequence from soybean 17.9 kDa HSP.

SOURCE: The coat protein gene was isolated from the potato leafroll virus luteovirus and multiplied in *E. coli*. The protein gene was mated into *Agrobacterium tumefaciens*. Russet Burbank potatoes were transformed with *Agrobacterium* containing the double gene constructs. Later products also contained the *cry3A* gene to introduce resistance to Colorado beetle (*Leptinotarsa decemlineata* (Say.)).

PRODUCTION: The transformed potato plants were used as parents in classical breeding programmes. 'NewLeaf Plus' is a Russet Burbank variety.

TARGET PESTS: Potato leafroll virus luteovirus and Colorado beetle.

TARGET CROPS: Potatoes.

BIOLOGICAL ACTIVITY: It has been demonstrated that the presence of the coat protein from viruses reduces or eliminates the symptoms of the virus.
Mode of action: Host resistance genes against potato leafroll luteovirus are polygenic, and this makes conventional breeding for resistance difficult. Transformed crops expressing the coat protein gene, however, show good resistance to attack from this aphid-transmitted virus. However, the induction of resistance to PLRV infection by the gene is not clearly understood at present. The gene does, however, induce virus resistance by a non-toxic mode of action. **Efficacy:** Good resistance has been shown to potato leafroll luteovirus.
Key reference(s): 1) GG Presting, OP Smith & CR Brown. 1995. Resistance to potato leafroll virus in potato plants transformed with the coat protein gene or with vector control constructs, *Phytopathol.*, **85**, 436–42. 2) D Duncan, D Hammond, J Zalewski, J Cudnohufsky, W Kaniewski, M Thornton, J Bookout, P Lavrik, G Rogan & JF Riebe. 1999. Field performance of 'Transgenic' potato with resistance to Colorado beetle and viruses. *HortScience*, **34(3)**, 556–7.

COMMERCIALISATION: Formulation: Sold as a component of potatoes transformed to show resistance to PLRV. **Trade names:** 'NewLeaf ' (*plv*), 'NewLeaf Plus' (*cry3A* plus *plv*) and 'NewLeaf Superior' (*cry3A* plus *plv* plus *epsps*) (Monsanto).

APPLICATION: 'NewLeaf' potatoes show resistance to PLRV.

PRODUCT SPECIFICATIONS: The transformed potato contains 0.03% PLRV resistance gene.

COMPATIBILITY: Compatible with all compounds registered for use in potatoes.

MAMMALIAN TOXICITY: 'NewLeaf' potatoes are considered not to be substantially different from conventionally bred potatoes and, as such, are not considered to be hazardous to mammals. There are no records of allergic or other adverse toxicological effects from researchers, breeders or growers of the transgenic varieties. Considered to be non-toxic.

ENVIRONMENTAL IMPACT AND NON-TARGET TOXICITY: As the transgenic crop is considered not to be substantially different from conventionally grown crops, it is considered that it will not show any adverse effects on non-target organisms or on the environment.

DATE(S) OF REGISTRATION: A change in emphasis by Monsanto has led to all 'NewLeaf' varieties being withdrawn in 2001, although the US and Canadian registrations are still valid.

1.1.5 *potato virus Y coat protein* gene

Introduces resistance to potato virus Y potyvirus

NOMENCLATURE: Approved name: *potato virus Y potyvirus coat protein* gene.

Other names: *PVY coat protein* gene.

Development code(s): RBMT15-101, SEMT15-02, SEMT15-15 ('NewLeaf Y').

Promoter(s): Viral coat protein (potato potyvirus Y strain O (common strain)) – figwort mosaic virus (FMV) 35S; leader sequence from the *Glycine max* heat-shock protein, Hsp 17.9; *cry3A* – arabSSU1A: *A. thaliana* ribulose-1,5-bisphosphate carboxylase (Rubisco) small subunit promoter.

Trait introduction: The Russet Burbank cultivar, RBMT15-101, and Shepody potato cultivars SEMT15-02 and SEMT15-15 ('NewLeaf Y') were bioengineered to be resistant to two important potato pests, the Colorado beetle (*Leptinotarsa decemlineata* Say.) and the ordinary (O) strain of potato potyvirus Y (PVY-O).

SOURCE: The coat protein gene from potato virus Y potyvirus was isolated from potato virus Y potyvirus and was multiplied in *E. coli*. It is driven by the promoter isolated from figwort mosaic virus.

PRODUCTION: The gene is transferred into a breeding line of potatoes and this is used as a parent in the development of virus-resistant strains of potato. 'NewLeaf Y' is bred from Russet Burbank and Shepody varieties. **Events:** 'NewLeaf Plus' RBMT21-129 and 21-350. 'NewLeaf Superior' RBMT22-88.

TARGET PESTS: Potato virus Y potyvirus.

TARGET CROPS: Potatoes.

BIOLOGICAL ACTIVITY: It has been demonstrated that the presence of the coat protein from viruses reduces or eliminates the symptoms of the virus.
Mode of action: Two of the most important viruses of potatoes are potato virus Y (PVY) potyvirus (aphid-transmitted) and potato virus X (PVX) potyvirus (mechanically transmitted). Mixed populations of these viruses occur frequently in cultivated potato crops and, together, they produce a synergistic increase in disease severity. The control of PVY, therefore, is effective in reducing the damage that results from the single virus and the combination.
Efficacy: It has been shown that crops transformed with the coat protein of PVY are protected from many strains of the virus and suffer reduced symptoms when exposed to PVY and potato virus X. Sold in crops also expressing the endotoxin gene of *Btt* for protection against Colorado beetle (*Leptinotarsa decemlineata* (Say)).
Key reference(s): RN Beachy, S Loesch-Fries & NE Turner. 1990. Coat protein-mediated resistance against virus infection, *Ann. Rev. Phytopathol.*, **28**, 451–74.

COMMERCIALISATION: Transgenic potatoes expressing the PVY coat protein gene and the endotoxin gene from *Btt* have been commercialised. **Trade names:** 'NewLeaf Y' (*pvy* and *cry3A*) (Monsanto).

APPLICATION: 'NewLeaf Y' potatoes show resistance to PVY and to Colorado beetle (*L. decemlineata*). For this reason, less insecticide has to be used to control aphids and leaf-feeding beetles.

PRODUCT SPECIFICATIONS: The 'NewLeaf Plus' potato also contains the *Bt*-toxin gene. Potato lines RBMT15-101, SEMT15-02 and SEMT15-15 were field tested in the USA (1994–98) and Canada. Field evaluations determined that vegetative vigour, overwintering capacity, insect and disease resistance (except for resistance to Colorado beetle and PVY), tuber yield and quality characteristics remained unchanged in comparison to commercial varieties.

COMPATIBILITY: Compatible with all crop protection agents that are registered for use on potatoes.

MAMMALIAN TOXICITY: 'NewLeaf Y' potatoes are considered not to be substantially different from conventionally bred potatoes and, as such, are not considered to be hazardous to mammals. There are no records of allergic or other adverse toxicological effects from researchers, breeders or growers of the transgenic varieties. Considered to be non-toxic.

ENVIRONMENTAL IMPACT AND NON-TARGET TOXICITY: As the transgenic crop is considered not to be substantially different from conventionally grown crops, it is considered that it will not show any adverse effects on non-target organisms or on the environment.

DATE(S) OF REGISTRATION: 'NewLeaf Plus' was registered in 1998 in the USA and in 1999 in Canada. 'NewLeaf Atlantic' and 'NewLeaf Superior' were registered in the USA in 1999 and in Canada in 2000. A change in emphasis by Monsanto has led to all 'NewLeaf' varieties being withdrawn in 2001, although the US and Canadian registrations are still valid.

1.1.6 *squash virus coat protein* genes

Introduces resistance to virus

NOMENCLATURE: Approved name: *squash virus coat protein* genes; cucumber mosaic virus cucumovirus (CMV) plus zucchini yellows mosaic potyvirus (ZYMV) plus water melon mosaic virus 2 potyvirus (WMV2) coat proteins.

Other names: CMV plus ZYMV plus WMV2 coat protein (CP) genes.

Development code(s): CZW-3; ZW20.

Promoter(s): Cauliflower mosaic virus (CaMV) 35S.

Trait introduction: *Agrobacterium tumefaciens*-mediated plant transformation. The yellow crookneck squash (*Curcurbita pepo* L.), line CZW-3 was developed using recombinant DNA techniques to be resistant to infection by three plant viruses, cucumber mosaic cucumovirus (CMV), zucchini yellow mosaic potyvirus (ZYMV) and watermelon mosaic potyvirus (WMV2). The novel variety was developed by insertion of the coat protein (CP) encoding sequences from these three single-stranded RNA viruses.

SOURCE: The coat protein genes were isolated from the three viruses that are important in the culture of cucurbits and replicated in *E. coli*. These genes were then used to transform squash plants. The promoter was isolated from the cucumber mosaic virus and has been found to be very effective at enhancing transcription levels of foreign genes in plants.

PRODUCTION: The transformants were used as parents in classical breeding programmes. ZW20 squash was used as one of the parents of CZW-3 ('Freedom II').

TARGET PESTS: Cucumber mosaic virus cucumovirus (CMV), zucchini yellows mosaic potyvirus (ZYMV) and water melon mosaic virus 2 potyvirus (WMV2).

TARGET CROPS: Squash.

BIOLOGICAL ACTIVITY: It has been demonstrated that the presence of the coat protein from viruses reduces or eliminates the symptoms of the virus. Although the rationale behind the creation of lines CZW-3 and ZW20 was to express the viral coat protein in order to interfere with the regulation of the viral infection (pathogen-derived resistance), the resistant plants exhibited remarkably low levels of the viral protein. It is most likely therefore, that the mechanism behind their virus-resistance is RNA silencing.
Mode of action: The search for a virus-resistant transformant examined the incorporation of genes coding for the coat proteins of CMV, ZYMV and/or WMV2. One of these lines was coded ZW-20, and it showed good resistance to mixed infections of ZYMV and WMV2. A transgenic squash hybrid was developed from ZW-20, and this became the first virus-resistant genetically engineered crop to be deregulated by USDA-ARS. **Efficacy:** The transgenic varieties have been shown to be resistant to attack from the three major cucurbit viruses and also produced excellent quality fruit. **Key reference(s):** 1) J P Arce-Ochoa, F Dainello, L M Pike & D Drews. 1995. Field performance comparison of two transgenic summer squash hybrids to their parental hybrid line, *HortScience*, **30**, 492–3. 2) G H Clough & P B Hamm. 1995. Coat protein transgenic resistance to watermelon mosaic and zucchini yellows mosaic virus in squash and canteloupe, *Plant Disease*, **29**, 1107–9.

COMMERCIALISATION: Commercialised by Asgrow Seed, following development work at Cornell University. **Trade names:** 'Freedom II Squash' (Asgrow Seed) and (Seminis Vegetable).

PRODUCT SPECIFICATIONS: The expression of viral CPs in transgenic squash varieties does not result in the formation of any infectious particles, or in the manifestation of any disease pathology, but rather enables the plant to resist infection by CMV, ZYMV and WMV2.

COMPATIBILITY: Compatible with all compounds recommended for use in squash.

MAMMALIAN TOXICITY: 'Freedom II Squash' is considered not to be substantially different from conventionally bred squash and, as such, is not considered to be hazardous to mammals. There are no records of allergic or other adverse toxicological effects from researchers, breeders or growers of the transgenic varieties. The CMV, WMV2 and ZYMV coat protein sequences were compared to databases of known protein toxins and did not show homologies with known mammalian protein toxins. The history of known safe consumption of these proteins from virus-infected plant products provides additional evidence of lack of toxicity. Considered to be non-toxic.

ENVIRONMENTAL IMPACT AND NON-TARGET TOXICITY: As the transgenic crop is considered not to be substantially different from conventionally grown crops, it is considered that it will not show any adverse effects on non-target organisms or on the environment. The transgenic squash line CZW-3 was field tested in the USA (1993–96). CZW-3 was extensively tested in laboratory, glasshouse and field experiments, and based on these assessments it was determined that the plants exhibited the typical agronomic characteristics of the parent crookneck squash, with the addition of resistance to CMV, ZYMV and WMV2 infection. Field trial reports demonstrated that CZW-3 did not exhibit weedy characteristics, and had no effect on non-target organisms or the general environment.

DATE(S) OF REGISTRATION: Regulatory approvals for CZW-3 for food and/or feed uses were granted in the USA in 1994 and environmental approval was received in 1996. Food use approvals were granted in Canada in 1998. Environmental regulatory approvals for ZW20 were granted in the USA in 1994; food and/or feed approvals were received in the USA in 1997; food approvals were granted in Canada in 1998.

1.2 Herbicide tolerance

1.2.1 *bromoxynil tolerance* gene

Introduces tolerance to bromoxynil

NOMENCLATURE: Approved name: *bromoxynil tolerance* gene.

Other names: *bxn* gene; nitrilase gene.

Development code(s): Canola – ACS-BNØ11-5 (OXY-235). Cotton – 31807 and 31808; and BXN.

Promoter(s): OXY-235 – CaMV 35S – Enhancer: non-translated leader of Rubisco small subunit gene derived from maize; 31807 and 31808 – chimera of the 35S promoter and the promoter from a mannopine synthase gene from *A. tumefaciens*; *cry1A*(*c*) – CaMV 35S.

Trait introduction: OXY-235 – *Agrobacterium tumefaciens*-mediated plant transformation. Cotton lines 31807 and 31808 were produced by *Agrobacterium*-mediated transformation in which the transfer-DNA (T-DNA) region of the bacterial tumour-inducing (Ti) plasmid was modified to contain DNA sequences encoding Cry1Ac protein and the nitrilase enzyme.

SOURCE: The gene used to transform cotton, canola and tobacco was isolated from the soil bacterium, *Klebsiella ozaenae* (Abel) Bergey *et al.*

PRODUCTION: The gene was characterised in and isolated from the originating organism, *K. ozaenae*, and then amplified in bacterial culture. Cotton callus was transformed using *A. tumefaciens* and transformants grown on in a collaborative programme between Rhône-Poulenc Agro (now Bayer CropScience) and Calgene. Canola was transformed using a disarmed non-pathogenic *A. tumefaciens* vector; the vector contained the transfer DNA (T-DNA) region of an *A. tumefaciens* plasmid from which virulence- and disease-causing genes had been removed, and replaced with the bromoxynil tolerance gene. The T-DNA portion of *A. tumefaciens* plasmids is generally known to insert randomly into the plant's genome and the insertion is usually stable, as it was shown to be in this case (event Oxy-235). Rhône-Poulenc has provided information on segregation and from Southern blot analysis that demonstrates that Oxy-235 has a single genetic insert, consisting of a single copy of the bromoxynil tolerance gene. The transformed plants were then used as parents in a breeding programme and elite varieties were selected for commercialisation.

TARGET PESTS: Hard-to-kill broad-leaved weeds, such as morning glory (*Ipomoea* spp.) and cocklebur (*Xanthium* spp.).

TARGET CROPS: Cotton, canola and tobacco.

BIOLOGICAL ACTIVITY: The *bxn* gene codes for a nitrilase enzyme that detoxifies bromoxynil, converting it to the non-herbicidal metabolite, dibromohydroxybenzoic acid. **Biology:** Preliminary work on *bxn*-tobacco, conducted by Rhône-Poulenc working with the French tobacco company SEITA, showed that transformed tobacco, known as dark 'ITB', would tolerate applied bromoxynil. This tobacco was granted a licence in Europe in March 1994 and was the first transgenic crop to achieve this status. Rhône-Poulenc worked closely with Calgene on the development of *bxn*-cotton and this was deregulated by USDA in February 1994. It is expected that the gene will be introduced into other crops, such as oilseed rape. **Mode of action:** Transformed crops express the gene constitutively and are, therefore, tolerant of bromoxynil applied over-the-top at recommended rates. **Efficacy:** Very effective at controlling hard-to-kill broad-leaved weeds without damage to the crop.

COMMERCIALISATION: Formulation: Introduced into US cotton-growing regions in 1998 and into canola-growing areas in North America in 1999. **Trade names:** 'BXN-Cotton' (Stoneville Pedigreed Seed) and (Calgene) (withdrawn), 'Bt plus Buctril BXN System Cotton' (*cryIA(c)* plus *bxn*) (31807 and 31808) (Monsanto) and (Stoneville Pedigreed Seed) (withdrawn), 'BXN-Canola' and 'Westar-Oxy-235' (Aventis) (withdrawn).

PRODUCT SPECIFICATIONS: The gene is expressed at a level of 1 g nitrilase per milligram of total protein in leaves and approximately 100-fold less in seeds. The Oxy-235 line was field tested in Canada from 1992 to 1996 and assessed for a number of agronomic characteristics, including seed yield, days to maturity, silique shattering and overwintering capacity. In comparison with non-transgenic commercial canola varieties, all of these parameters were within the normal range of expression. Stress adaptation was evaluated, including resistance to major canola pests such as the fungal pathogen blackleg (*Leptosphaeria maculans* (Desm.) Ces. & de Not.), and the susceptibility of Oxy-235 was within the ranges currently displayed by commercial varieties. The only significant difference between Oxy-235 canola and the parental non-transformed variety was that Oxy-235 exhibited field tolerance to ioxynil and bromoxynil. The transgenic cotton lines 31807 and 31808 were field tested in the USA, and it was concluded that they did not exhibit weedy characteristics and had no effect on non-target organisms or the general environment.

COMPATIBILITY: Compatible with a wide variety of crop protection chemicals used in cotton cultivation.

MAMMALIAN TOXICITY: The nitrilase protein showed no significant homology with any known toxins or allergens in the Genebank and Swiss-Pro databases. The nitrilase protein was also shown to have low stability under simulated gastric conditions. The enzyme that confers bromoxynil tolerance is a bacterial version of an enzyme that is ubiquitous in Nature, including monocotyledonous plants such as maize, wheat and barley, and, therefore, would not be expected to be toxic or allergenic.

ENVIRONMENTAL IMPACT AND NON-TARGET TOXICITY: *Bxn*-cotton and canola have not been shown to have any adverse effects on non-target organisms or on the environment.

DATE(S) OF REGISTRATION: Regulatory approvals for OXY-235 for the environment were granted in Canada in 1997 and Japan in 1998; food use approval was granted in Canada in 1997, in Japan and the USA in 1999 and in Australia in 2002; feed use approval was given in Canada in 1997 and in Japan in 1999. Environmental regulatory approvals for cotton lines 31807 and 31808 were granted in the USA in 1997 and Japan in 1998; food and/or feed approval was granted in the USA in 1998; food use approval was given in Canada in 1998 and Japan in 1999; and feed uses were approved in Japan in 1999. Environmental regulatory approvals for *bxn*-cotton were granted in the USA in 1994 and in Japan in 1997; food and/or feed use approvals were granted in the USA in 1994, Mexico in 1996 and Australia in 2002; food uses were approved in Canada in 1996 and Japan in 1997; and feed uses were given in Canada in 1996 and in Japan in 1998.

1.2.2 *csr1-2* gene

Introduces tolerance to imidazolinone herbicides

NOMENCLATURE: Approved name: *csr1-2* gene, *ahasl* gene.

Development code(s): CV127; BPS-CV127-9.

Promoter(s): Not relevant as selected rather than introduced.

Trait introduction: A purified, linear DNA fragment derived from plasmid pAC321 was used to transform embryogenic axis tissue derived from the apical meristem of a single soybean seed of the commercial variety 'Conquista' using particle bombardment. No carrier DNA was used in the process. Soybean tissues were transformed with an approximately 6.2 kb linear fragment *PvuII* fragment derived from plasmid pAC321 containing the *csr1-2* gene cassette (also referred to in the literature as *ahasl*).

SOURCE: CV127 soybean has been genetically modified to express an altered AtAHASL protein that is encoded by the *csr1-2* gene from *Arabidopsis thaliana* and confers tolerance to imidazolinone herbicides. The AtAHASL protein encoded by *csr1-2* is structurally and functionally identical to the native AtAHASL, except for substitution of a serine with an asparagine at residue 653 (S653N), which results in tolerance to imidazolinone herbicides.

TARGET PESTS: A wide range of weeds.

TARGET CROPS: Soybean.

BIOLOGICAL ACTIVITY: CV127 soybeans show increased tolerance to imidazolinone herbicides over conventional varieties.

COMMERCIALISATION: Trade names: 'CV127 soybean' (*csr1-2*) (BASF).

COMPATIBILITY: Compatible with all imidazolinone herbicides and herbicides generally used to control weeds in maize. Volunteer crops can be controlled by the use of herbicides that are used to remove conventional varieties, with the exception of imidazolinone herbicides.

MAMMALIAN TOXICITY: CV127 soybean plants have been shown to be substantially identical to related varieties and are not expected to show any adverse toxicological effects. There have been no reports of allergic or other adverse toxicological effects from research workers, breeders or growers.

ENVIRONMENTAL IMPACT AND NON-TARGET TOXICITY: CV127 soybean plants are not expected to have any adverse effects on non-target organisms or on the environment.

DATE(S) OF REGISTRATION: CV127 soybean plants were approved in Brazil for the environment and for use as food and/or feed in 2009.

1.2.3 *cp4 epsps*

Introduces tolerance to glyphosate

NOMENCLATURE: Approved name: *cp4 epsps*.

Other names: Glyphosate-tolerant form of 5-enolpyruvylshikimate-3-phosphate synthase.

Development code(s): *Agrostis stolonifera* L. (creeping bentgrass) – ARS368; sugar beet – GTSB77 and H7-1; canola – MON89249-2 (GT200), MON-ØØØ73-7 (GT73, RT73) and ZSR500/502; soybean – MON-Ø4Ø32-6 (GTS 40-3-2) and MON-89788-1; cotton – DAS-21Ø23-5 × DAS-24236-5 × MON-Ø1445-2, DAS-21Ø23-5 × DAS-24236-5 × MON88913, GHB614, MON-15985-7 × MON-Ø1445-2, MON-ØØ531-6 × MON-Ø1445-2, MON1445/1698, MON15985 × MON88913, MON 88913; alfalfa (lucerne) – J101 and J163; potato – RBMT21-129, RBMT21-350, RBMT22-082; wheat – MON71800; maize (corn) – GA21, BT11 × GA21, BT11 × MIR604 × GA21, NK603, BT11 × MIR604 × GA21, DAS-59122-7 × NK603, MON80100 × MON802, MON809, MON810 × MON88017,

MON832, MON863 × MON810 × NK603, MON863 × NK603, MON88017, MON89034 × MON88017, MON89034 × NK603, MON89034 × TC1507 × MON88017 × DAS-59122-7, TC1507 × NK603, NK603 × T25, NK603 × MON810, DAS-59122-7 × TC1507 × NK603.

Promoter(s): *A. stolonifera* – enhanced CaMV 35S, HSP70 intron, gox cassette chloroplast transit peptide from *Arabidopsis thaliana epsps* gene (CTP2) and rice actin I promoter and intron sequences chloroplast transit peptide from *A. thaliana epsps* gene (CTP2); sugar beet GTSB77 and H7-1 (*cp4 epsps*) – figwort mosaic virus (FMV) 35S chloroplast transit peptide from *A. thaliana*; H7-1 (*goxv247*) – figwort mosaic virus (FMV) 35S chloroplast transit peptide from *A. thaliana* (upon insertion into the plant genome, the GOX-encoding gene was truncated and fused to genomic sugar beet DNA sequences, resulting in a chimeric sequence that did not express a functional protein); canola GT200, GT73, RT73, ZSR500/502 (*cp4 epsps* and *goxv247*) – chloroplast transit peptide from *A. thaliana*; soybean GTS 40-3-2 (*cp4 epsps*) – enhanced CaMV 35S chloroplast transit peptide from *Petunia hybrida*; MON-89788-1 – P-FMV/TSF1: chimeric promoter containing 008 *A. thaliana* tsf1 promoter elongation factor EF-1 alpha; figwort mosaic virus 35S promoter enhancer; *A. thaliana* tsf1 leader and intron; chloroplast transit peptide from *A. thaliana*. Cotton – 'WideStrike' × MON88913 *cp4 epsps* – P-35S/ACT8: chimeric promoter from *A. thaliana* act8 f cauliflower mosaic virus 35S promoter enhancer; *A. thaliana* act8 leader and intron; chloroplast transit peptide from *A. thaliana*; and P-FMV/TSF1: chimeric promoter containing *A. thaliana* tsf1 promoter elongation factor EF-1 alpha; figwort mosaic virus 35S promoter enhancer; *A. thaliana* tsf1 leader and intron; chloroplast transit peptide from *A. thaliana*; *cry1Ac* – ubiquitin 1 (*Zea mays*); *cry1F* – mannopine synthase (d mas 2') promoter from pTi15955 four copies of the octopine synthase (4OCS) enhancer from pTiAch5; *pat* – ubiquitin (ubi) ZM (*Z. mays*) promoter and the first exon and intron. DAS-21Ø23-5 × DAS-24236-5 × MON-Ø1445-2 *CP4 epsps* – CMoVb from modified figwort mosaic virus (FMV) chloroplast transit peptide from *A. thaliana epsps* gene (CTP2), *cry1Ac* – ubiquitin 1 (*Z. mays*) and *cry1F* – mannopine synthase (d mas 2') promoter from pTi15955 four copies of the octopine synthase (4OCS) enhancer from pTiAch5; *pat* – ubiquitin (ubi) ZM (*Z. mays*) promoter and the first exon and intron. GHB614 – promoter region of the histone H4 gene from *A. thaliana*. Intron of gene II of the histone H3.III variant from *A. thaliana*. MON 88913 – *epsps* – P-FMV/TSF1: chimeric promoter containing *A. thaliana* tsf1 promoter elongation factor EF-1 alpha; figwort mosaic virus 35S promoter enhancer; *A. thaliana* tsf1 leader and intron; chloroplast transit peptide from *A. thaliana* and P-35S/ACT8: chimeric promoter from *A. thaliana* act8 f cauliflower mosaic virus 35S promoter enhancer plus *A. thaliana* act8 leader and intron; chloroplast transit peptide from *A. thaliana*; MON71800 – 2 × *cp4 epsps* – enhanced CaMV 35S chloroplast transit peptide from *A. thaliana epsps* gene (CTP2) and rice actin I promoter and intron sequences chloroplast transit peptide from *A. thaliana* EPSPS gene (CTP2); GA21 – *epsps* – rice actin I promoter and intron sequences; ribulose-1,5-bisphosphate carboxylase oxygenase (Rubisco)-derived chloroplast transit peptide sequences (from corn and sunflower);

Trait introduction: ASR368 creeping bentgrass was produced by microprojectile bombardment of plant cells from the creeping bentgrass line B99061R. Sugar beet lines GTSB77 and H7-1 were produced by *Agrobacterium*-mediated transformation of the proprietary cytoplasmic male sterile sugar beet line A1012 with plasmid PV-BVGT03 (upon insertion into the plant genome, the GOX-encoding gene of GTSB77 was truncated and fused to genomic sugar beet DNA sequences, resulting in a chimeric sequence that did not express a functional protein). Unless otherwise listed, all glyphosate-tolerant crops containing the *cp4 epsps* and the *goxv247* genes were transformed by a disarmed non-pathogenic *A. tumefaciens* vector; the vector contained the transfer DNA (T-DNA) region of an *A. tumefaciens* plasmid, from which virulence and disease causing genes were removed and replaced with the 'Roundup Ready' genes. GTS 40-3-2 was developed by introducing the CP4 EPSPS coding sequence into the soybean variety A5403, a commercial soybean variety of Asgrow Seed Company, using particle-acceleration (biolistic) transformation. A5403 is a maturity group V cultivar. MON89788 was developed by introducing the *CP4 epsps* gene into a commercial soybean variety using *A. tumefaciens*-mediated transformation of meristem-derived material from soybean variety A3244, an elite maturity group III soybean variety. Cotton – 'WideStrike' × MON1445 (DAS-21Ø23-5 × DAS-24236-5 × MON-Ø1445-2) was produced by crossing 'WideStrike' insect-resistant cotton (DAS-21Ø23-5 × DAS-24236-5) with the herbicide-tolerant cotton line MON1445. This stacked cotton line is a product of traditional plant breeding. Cotton line GHB614 was developed through *Agrobacterium*-mediated transformation of the cotton variety 'Coker' 312, using the transformation vector pTEM2. Cotton explants were exposed to a culture of disarmed *A. tumefaciens* containing plasmid p-TEM2. After co-culture, the cotton cells were regenerated to whole plants using the appropriate regeneration media with 500 mg/L claforan to eliminate residual *Agrobacterium* and then selected with glyphosate. MON-15985-7 × MON-Ø1445-2 cotton is the product of the cross-breeding of the insect-resistant cotton line 15985 (MON-15985-7) with the herbicide-tolerant cotton line MON1445. MON15985 × MON88913 cotton is the product of the cross-breeding of the insect-resistant cotton line 15985 (MON-15985-7) with the herbicide-tolerant cotton line MON 88913. MON 88913 cotton was developed by introducing two CP4 EPSPS coding sequences into the cotton variety 'Coker 312' using *Agrobacterium*-mediated transformation. Events J101 and J163 were each developed by introducing the CP4 EPSPS coding sequences into the alfalfa clone 'R2336' using *Agrobacterium*-mediated transformation. MON 71800 was developed by introducing the CP4 EPSPS coding sequences into the spring wheat variety 'Bobwhite' using *Agrobacterium*-mediated transformation. BT11 × GA21 10 maize is an F1 hybrid resulting from the hybridisation of the lepidopteran-resistant and herbicide-tolerant maize line BT11 (SYN-BTØ11-1) and the herbicide-tolerant maize line GA21 (MON-ØØØ21-9) and is a product of traditional plant breeding. BT11 × MIR604 × GA21 maize is an F1 hybrid resulting from the hybridisation of the lepidopteran-resistant and glufosinate-tolerant maize line BT11 (SYN-BTØ11-1), the coleopteran-resistant maize line MIR604 (SYN-IR6Ø4-5) and the glyphosate-tolerant maize line GA21 (MON-ØØØ21-9). This stacked maize hybrid is a product of traditional plant breeding. The GA21 line of maize (*Zea mays* L.) was genetically engineered, by particle acceleration (biolistic) transformation, to be tolerant of glyphosate-containing herbicides. The isolated endogenous maize 5-enolpyruvylshikimate-3-phosphate synthase (*epsps*) gene was modified through site-directed mutagenesis, such that its encoded enzyme was insensitive to inactivation by glyphosate, and inserted into the inbred AT maize variety.

SOURCE: The *cp4 epsps* gene was isolated from *A. tumefaciens* strain CP4 isolated from a glyphosate production facility and cloned into *E. coli*.

TARGET PESTS: A wide variety of weeds (post-emergence applications).

TARGET CROPS: ARS368 is to be used in the production of turf in golf courses, with an incidental usage in livestock feed. Soybean, cotton, wheat, lucerne, sugar beet and maize (corn).

BIOLOGICAL ACTIVITY: The *cp4 epsps* gene was isolated from *A. tumefaciens* strain CP4. The EPSPS protein is a chloroplast-localised enzyme that is transported from the cytosol to the chloroplast by the chlorophyll transit peptide. In transformed crops, an inserted gene codes for a chlorophyll transit peptide that binds to the CP4 EPSPS enzyme to enable transportation to the chloroplast.
Biology: The EPSPS enzyme is part of the shikimate pathway that is involved in the production of aromatic amino acids and other aromatic compounds in plants. When conventional plants are treated with glyphosate, the herbicide binds to EPSPS, thereby preventing the synthesis of aromatic amino acids needed for plant growth. The CP4 EPSPS enzyme has a reduced affinity for glyphosate; its enzymatic activity is, therefore, not inhibited by the herbicide. **Mode of action:** The advantage of the CP4 enzyme is that it shows a similar affinity to phosphoenol-pyruvate (PEP), the compound that reacts with shikimate-3-phosphate under the influence of EPSPS to form 5-enolpyruvoylshikimate-3-phosphate, as does the wild type, but it has significantly reduced affinity for glyphosate (2000 times less).
Efficacy: The recommended use rate for glyphosate is 0.34–1.12 kg a.e./hectare. At these rates, all weeds infesting transformed crops are well controlled, with no adverse effects on the crop. If the application is timed to coincide with small weed size and just prior to crop canopy cover, lower rates give excellent control.
Key reference(s): 1) G Kishore *et al.* 1988. EPSP synthase: from biochemistry to genetic engineering of glyphosate tolerance. In *Biotechnology for Crop Protection*, ACS Symposium Series No. 379, PA Hedin, JJ Menn & RM Hollingworth (eds.), pp. 37–48,

American Chemical Society, Washington, DC, USA. 2) JP Giesy, S Dodson & KR Solomon. 2000. Ecotoxicological risk assessment for Roundup herbicide. *Reviews of Environmental Contamination and Toxicology*, **167**, 35–120. 3) X Delannay, T Bauman, D Beighley, M Buettner, H Coble, M DeFelice, C Derting, T Diedrick, J Griffin, E Hagood, F Hancock, S Hart, B LaVallee, M Loux, W Lueschen, K Matson, C Moots, E Murdock, A Nickell, M Owen, E Paschal II, L Prochaska, P Raymond, D Reynolds, W Rhodes, F Roeth, P Sparankle, L Tarochione, C Tinius, R Walker, L Wax, H Weigelt & S Padgette. 1995. Yield evaluation of glyphosate-tolerant soybean line after treatment with glyphosate. *Crop Science* **35(5)**, 1461–7. 4) DL Nida, KH Kolacz, RE Buehler, WR Deaton, WR Schuler, TA Armstrong, ML Taylor, CC Ebert, GJ Rogan, SR Padgette & RL Fuchs. 1996. Glyphosate-tolerant cotton: genetic characterisation and protein expression. *Journal of Agricultural and Food Chemistry* **44(7)**, 1960–6.

COMMERCIALISATION: Trade names: 'Roundup Ready Creeping Bentgrass' (*cp4 epsps*) (Scotts Seeds), 'InVigor'(*CP4 epsps* plus *goxv247*) (Monsanto) and (Syngenta), 'Roundup Ready Sugar Beet' (*cp4 epsps*) (Monsanto) and (KWS), 'Roundup Ready Canola' (*cp4 epsps* plus *goxv247*) (Monsanto), 'Westar Roundup Ready Canola' (*cp4 epsps* plus *goxv247*) (Monsanto), 'Hysyn 101 Roundup Ready Canola' (*cp4 epsps* plus *goxv247*) (Monsanto), 'Roundup Ready Soybeans' (*cp4 epsps*) (Monsanto), 'Roundup Ready2Yield Soybean' (*cp4 epsps*) (Monsanto), 'WideStrike/Roundup Ready Cotton' (*cp4 epsps* plus *pat* plus *cry1A(c)* plus *cry1F*) (Dow AgroSciences (Mycogen)), 'Roundup Ready Flex Cotton' (*cp4 epsps*) (Dow AgroSciences) and (Pioneer), 'Glytol Cotton' (2 × *cp4 epsps* genes) (Bayer), 'Roundup Ready BollGard II Cotton' (*cp4 epsps* plus *cry1A(c)* plus *cryIIA(b2)*) (MON-15985-7 × MON-Ø1445-2) (Monsanto), 'Roundup Ready BollGard Cotton' (*cp4 epsps* plus *cry1A(c)*) (MON-ØØ531-6 × MON-Ø1445-2) (Monsanto), 'Roundup Ready Cotton' (*cp4 epsps*) (MON1445/1698) (Monsanto), 'Roundup Ready Flex Bollgard II Cotton' (*cp4 epsps* plus *cry1Ac* plus *cry2Ab*) (MON15985 × MON88913) (Monsanto), 'Roundup Ready Flex Cotton' (2 × *cp4 epsps*) (MON 88913) (Monsanto), 'Roundup Ready Alfalfa'(*cp4 epsps*) (J101 and J163) (Monsanto) and (Forage Genetics), 'Russet Burbank NewLeaf Plus' (*cry3A* plus *cp4 epsps* plus *hel*) (Monsanto) (withdrawn), 'Roundup Ready Wheat' (2 × *cp4 epsps*) (MON 71800) (Monsanto) (not commercialised), 'Agrisure GT/CB/LL' (*cry1A(b)* plus *cry3A* plus *cp4 epsps* plus *pat*) (BT11 × GA21) (Syngenta), 'YieldGard Corn Borer Corn with Roundup Ready 2' (*cry1A(b)* plus *epsps* (Bt MON 810 × NK603) (Monsanto), 'YieldGard Plus with Roundup Ready Corn 2' (*cry1A(b)* plus *cry3B(b1)* plus *cp4 epsps* (Bt MON 810 × MON 863 × NK603) (Monsanto), 'YieldGard VT Triple Rootworm Roundup Ready 2 Corn' (*cry1A(b)* plus *cry3B(b1)* plus *cp4 epsps*) (Bt MON 810 × MON 88017) (Monsanto), 'Agrisure 3000GT' (*cry1A(b)* plus *cry3A* plus *pat* plus *cp4 epsps*) (BT11 × MIR604) (Syngenta), 'Genuity VT' and 'Double PRO' (*cry1A.105* plus *cry2A(b2)* plus *cp4 epsps*) (MON 89034 × NK603) (Monsanto), 'Genuity VT Triple PRO' (*cry1A.105* plus *cry2A(b2)* plus *cry3B(b1)* plus *cp4 epsps*) (MON 89034 × MON 88017) (Monsanto),

'Genuity' and 'SmartStax' (*cry1A.105* plus *cry2A(b2)* plus *cry1F* plus *cry3B(b1)* plus *cry34A(b1)* plus *cry35A(b1)* plus *pat* plus *cp4 epsps*) (Monsanto) and (Dow AgroSciences), 'Herculex 1 plus Roundup Ready Corn 2' (*cry1F* plus *pat* plus *cp4 epsps*) (TC1507 plus NK603 (crossed)) (Dow AgroSciences), (Mycogen), (Monsanto) and (Pioneer), 'Herculex Xtra with Roundup Ready Corn 2' (*cry34A(b)1* plus *cry35A(b)1* plus *cry1F* plus *pat* plus *cp4 epsps*) (TC1507 plus DAS 59122-7 plus NK603 (crossed)) (Dow AgroSciences), (Monsanto), (Pioneer) and (DuPont), 'Roundup Ready Corn'(*cp4 epsps*) (GA21) (Syngenta Seeds), 'Roundup Ready Corn' (2 × *cp4 epsps*) (NK603) (Monsanto), 'Roundup Ready Corn 2'(*cp4 epsps* plus *goxv247*) (MON832) (Monsanto). **Patents:** US Patent 7335816 2008 (amongst others).

PRODUCT SPECIFICATIONS: The transgenic sugar beet line GTSB77 was field tested in the USA (1996–98). GTSB77 was evaluated extensively and agronomic data showed that there were no significant changes in the number of seeds produced, germination characteristics, final stand, overwintering capability or pathogen susceptibility. Sugar beet H7-1 was field tested in the USA from 1998 to 2002, and in Europe (France and Germany) in 1998–99. In addition to evaluation in different environments, H7-1 was also evaluated in different genetic backgrounds that represented regionally adapted cultivars. Field trial evaluations included agronomic performance, disease and insect resistance, leaf morphology, seed germination and dormancy, bolting, flowering onset and seed harvest date. In most field trials, H7-1 was compared to a control that represented a nearly equivalent genotype. No significant differences were observed in any of these characteristics between H7-1 and the non-transgenic controls. The canola line GT200 was field tested in Canada (1992–94). Agronomic characteristics such as vegetative vigour, overwintering capacity, time to maturity, seed production and yield of the transformed line GT200 were compared to unmodified canola counterparts and determined to be within the normal range of expression found in commercial canola cultivars. Stress adaptation was evaluated, including resistance to major canola pests such as the fungal pathogen blackleg (*Leptosphaeria maculans* (Desm.) Ces. & de Not.), and the susceptibility of GT200 was within the ranges currently displayed by commercial varieties. The only significant difference between GT200 canola and the parental non-transformed variety was tolerance to glyphosate. Overall, the field data reports demonstrated that canola GT200 had no potential to pose a plant pest risk. GTS 40-3-2 soybean has been field tested in the USA (1991–93), Canada (1992), Puerto Rico (1993), Argentina and Costa Rica. Agronomic studies on seed yield and visual observations on germination characteristics of seeds, final stands, disease and insect susceptibility supported the conclusion that soybean line GTS 40-3-2 was as safe to grow as other soybean varieties and had no potential to pose a plant pest risk. Field trials were conducted at 17 locations during 2005 to evaluate thoroughly phenotypic, agronomic and ecological interactions of MON89788 soybeans and compare these with A3244, the conventional parental line, and other commercially available soybeans. These 17 locations provided a diverse range of environmental and agronomic conditions representative of the majority of commercial soybean production regions in the USA. Plant growth stage was assessed several times during the growing season, and observational data on the presence of various biotic and abiotic stresses together with the response of the lines to these stresses was recorded. A total of 11 different phenotypic characters were evaluated and the only significant difference detected in across-site analyses was a reduced height for MON89788 compared with A3244. The difference falls within the range of values observed for commercial soybean varieties. No differences were observed in susceptibility to pests and diseases, neither were there any differences observed in numbers of beneficial insects and spiders collected from the sites. Seed dormancy and germination characteristics were determined across a range of conditions with seed produced during the 2005 trials at three different locations.

COMPATIBILITY: Compatible with a wide variety of crop protection chemicals used in crop cultivation.

MAMMALIAN TOXICITY: There is no evidence to indicate that the CP4-EPSPS protein has the potential to be either toxic or allergenic to humans. Proteins from the EPSPS family are naturally present in our food supply or expressed in human intestinal microflora. Furthermore, the protein was not detectable in the principal food components of the crops. The potential for toxicity and allergenicity of the CP4 EPSPS protein was assessed. An amino acid sequence homology search, using the ALLPEPTIDES database, revealed no similarities between CP4 EPSPS and known toxins. A sequence homology search conducted using the ALLERGEN3 database also revealed no similarities between the novel protein and known allergens, including gliadin. The CP4 EPSPS protein was degraded within one minute of exposure to simulated gastric and intestinal fluids, and the enzymatic activity reduced to less than 10% of the initial activity within 15 seconds of exposure. No adverse effects were observed in an acute toxicity study where mice were fed (by gavage) doses of CP4 EPSPS up to 572 mg/kg.

ENVIRONMENTAL IMPACT AND NON-TARGET TOXICITY: It has been concluded that the insertion of *CP4 epsps* into the transgenic crops will not result in any deleterious effects or significant impacts on non-target organisms, including threatened and endangered species or beneficial organisms. The enzyme EPSPS that confers glyphosate resistance is from a commonly occurring soil bacterium, *A. tumefaciens*, and is similar to the gene that is normally present in conventional crops and is not known to have any toxic properties. Field observations have revealed no negative effects on non-target organisms. The lack of known toxicity for this enzyme suggests no potential for deleterious effects on beneficial organisms such as bees and earthworms. The high specificity of the enzyme for its substrates makes it unlikely that the introduced enzyme would metabolise endogenous substrates to produce compounds toxic to beneficial organisms.

DATE(S) OF REGISTRATION: 'Roundup Ready Creeping Bentgrass' was approved for feed use in the USA in 2003. 'InVigor' sugar beet was approved for the environment and food and/or feed uses in the USA in 1998; it was approved for food and/or feed use in Australia in 2002; for food use in Japan in 2003; and for feed and food use in the Philippines in 2004. 'Roundup Ready Sugar Beet' was registered for food and/or feed use in the USA in 2004 and for the environment in 2005; in Canada approval was given for the environment and for food and feed uses in 2005; Australia granted food uses in 2005; Mexico allowed food and/or feed use in 2006; Japan approved food use in 2003, and feed uses and environmental clearance in 2007; Colombia and the EU granted food and/or feed uses in 2007. Regulatory approvals for 'Roundup Ready Canola' were granted in Canada for the environment in 1996 and for food and feed uses in 1997; Japan approved food and feed uses in 2001 and for the environment in 2006; in the USA food and/or feed uses were approved in 2002 and for the environment in 2003. Regulatory approvals for 'Westar Roundup Ready Canola' were granted in Canada for food uses in 1994 and for feed use and the environment in 1995; in the USA for food and/or feed in 1995 and for the environment in 1999; in Mexico for food and/or feed uses in 1996; in Japan for the environment, food and feed uses in 1996; in the EU for food use in 1997 and for feed use in 2005; in Australia for food use in 2000 and for the environment in 2003; in the Philippines for food and feed uses in 2003 and in Korea for food use in 2003 and feed uses in 2005. 'Hysyn 101 RR Roundup Ready Canola' was approved for the environment and for feed use in Canada in 1997. Regulatory approvals for GTS 40-3-2 soybeans have been granted in the USA in 1994 for the environment and food and/or feed uses; in Canada for the environment and feed uses in 1995 and food uses in 1996; in Japan and Argentina for the environment, food and feed uses and in Switzerland and the UK for food and feed use in 1996; in Uruguay for the environment, food and feed use in 1997; in Australia and Korea for food use in 2000; in South Africa for the environment, food and feed use and in the Czech Republic for food and feed uses in 2001; in Taiwan in 2002 for food use; in the Philippines for food and feed use in 2003; and in Korea for feed use in 2004. Regulatory approvals for 'Roundup Ready2Yield Soybean' were granted in 2007 in the USA for the environment and food and/or feed use and in Canada for the environment and for food and feed use; in Japan food use approval was given in 2007 and feed use in 2008; food and/or feed use was approved in the Philippines in 2007; food and/or feed use approvals were granted in Australia, China, the EU and Mexico in 2008; Korea gave approval for food and feed uses in 2009. 'WideStrike/Roundup Ready Cotton' received approval for food and/or feed in Mexico in 2005; for feed and food uses in Japan and food use in Korea in 2006. Regulatory approvals for 'WideStrike' × MON88913 cotton were granted for food and/or feed in Mexico, for food and feed use in Japan and for food use in Korea in 2006. 'GlyTol' regulatory approvals were granted for food and/or feed use in Mexico and the USA and for food and feed use in Canada in 2008, and environmental clearance was given in the USA and food use in Australia in 2009. MON-15985-7 × MON-Ø1445-2 received environmental regulatory approval in Australia in 2002; food use in Korea and the Philippines in 2004 food and/or feed approval in the EU and Japan in 2005 and in Mexico in 2006; and feed use approval in the Philippines in 2004 and Korea in 2008. Regulatory approvals for 'Roundup Ready Bollgard' cotton were granted for food and/or feed use in Mexico in 2002, in Japan in 2004, in the EU and South Africa in 2005 and in Brazil in 2009; environmental approval was granted in Australia in 2003, in South Africa in 2005 and in Argentina and Brazil in 2009; food use clearance was given in Korea and the Philippines in 2004, in Brazil in 2008 and Argentina in 2009; feed use was approved in the Philippines in 2004, Korea in 2008 and Argentina in 2009. Environmental regulatory approvals for 'Roundup Ready Cotton' were granted in the USA in 1995, Japan in 1997, Argentina in 1999, Australia and South Africa in 2000, Colombia in 2004 and Brazil in 2008; food and/or feed use approvals were granted in the USA in 1995, Mexico and South Africa in 2000, China and Colombia in 2004, the EU in 2005 and Brazil in 2008; food use clearance was granted in Canada in 1996, in Korea in 1997, in Australia in 2000, in Argentina in 2001 and in Korea and the Philippines in 2003; feed use approvals were agreed in Canada in 1996, in Japan in 1997, in Australia in 2000, in Argentina in 2001 and in Korea and the Philippines in 2003. Environmental regulatory approvals for 'Roundup Ready Flex Bollgard II Cotton' were granted in Australia in 2006 and South Africa in 2007; food and/or feed uses were approved in Mexico in 2006 and in South Africa in 2007; food uses were approved in Japan in 2005 and in Korea and the Philippines in 2006; feed use was approved in Japan and the Philippines in 2006, Colombia in 2007 and Korea in 2008. 'Roundup Ready Flex Cotton' was granted environmental regulatory approval in the USA in 2004, Australia in 2006 and in South Africa in 2007; food and/or feed approval in the USA and the Philippines in 2005, in Mexico in 2006 and in China, Colombia and South Africa in 2007; food approval was granted in Canada and Japan in 2005 and in Australia and Korea in 2006; use as feed was approved in Canada in 2005 and in Japan and Korea in 2006. Environmental regulatory approval was given for 'Roundup Ready Alfalfa' in Canada in 2005 and Japan in 2006; food and/or food use clearance was granted in the USA in 2004 and Mexico in 2005, food use clearance was achieved in Canada and Japan in 2005, in Australia and the Philippines in 2006 and in Korea in 2007; feed use approvals were granted in Canada in 2005, in Japan and the Philippines in 2006 and in Korea in 2007. Environmental regulatory approval for 'NewLeaf Plus' potato was granted in the USA in 1998 and Canada in 1999; food and/or feed use was granted in the USA in 1998 and in Australia and Mexico in 2001; food use was allowed in Canada in 1999 and Korea and the Philippines in 2004; and feed approval was granted in Canada in 1999, Japan in 2001 and the Philippines in 2004. 'Roundup Ready Wheat' was granted food and/or feed approval in Canada and the USA in 2004. Environmental regulatory approvals for 'Agrisure GT/CB/LL' were granted in Canada in 2005, Japan in 2007 and Brazil in 2009; food and/or feed uses were approved in Mexico in 2007 and Brazil in 2009; food use was approved in Korea in 2006 and in Japan and the Philippines in 2007; and feed use was approved in the Philippines in 2007 and Korea in 2008. An application for renewal of an existing product according to Regulation 1829/2003 on genetically modified food and feed was requested for MON-15985-7 × MON-Ø1445-2 (*crylA(c)* plus *cry2A(b2)* plus *cp4 epsps* genes). This was checked for completeness and accepted on 9 April 2008 by EFSA. However, additional information was requested from the applicant and the proceedings have been suspended. An application for renewal of an existing product according to Regulation 1829/2003 on genetically modified food and feed was made for MON531 × MON1445 (*crylA(c)* plus *cp4 epsps*) on 17 April 2007. This was checked for completeness and accepted on 12 March 2008, but additional information has been requested from the applicant and the proceedings have been suspended. An application for MON15985 011× MON88913 (*cryla(c)* plus *cry2a(b2)* plus *cp4 epsps*) was made in April 2007 and checked for completeness by the EFSA. Additional information has been requested from the applicant and the proceedings have been suspended. Food and/or feed approvals for MON89034 × MON88017 (*cry1A.105* plus *cry2Ab(2)* plus *cry3Bb(1)* plus *cp4 epsps*) were granted in Japan in 2008, in the Philippines in 2009 and in Mexico in 2010. It was approved for food use in Korea and Taiwan in 2009 and for use as feed in Taiwan in 2009. Regulatory approvals for MON89034 × NK603 (*cry1A.105* plus *cry2Ab* plus 2 × *cp4 epsps*) were granted for food and/or feed in Japan in 2008 and in Mexico in 2010, for food in the Philippines and Taiwan in 2009 and in Korea in 2010, and for use as feed in Korea and the Philippines in 2009. Regulatory approvals for MON89034 × TC1507 × MON88017 × DAS-59122-7 (*pat* plus *cp4 epsps* plus *cry1A.105* plus *cry2Ab* plus *cry3Bb1* plus *cry34Ab1* plus *cry35Ab1* plus *cry1Fa2*) were granted for environmental impact in Canada, Japan and the USA in 2009; for food and/or feed in Mexico and the Philippines in 2010; and for food and use as feed in Japan, Korea and Taiwan in 2009. Regulatory approvals for BT11 × MIR604 × GA21 (*pat* plus *cp4 epsps* plus *cry3A* plus *cry1Ab*) for the environment in Canada in 2007 and in Korea in 2008, for food and/or feed in Mexico in 2008, for food in Japan in 2007 and in Korea and the Philippines in 2008 and for use as feed in the Philippines in 2008. Regulatory approvals for GA21 × MON810 (*cry1Ab* plus *epsps*) were granted for food and/or feed in Japan and South Africa in 2003 and in the EU in 2005; for food in Korea and the Philippines in 2004 and for feed in the Philippines in 2004. It was notified in the EU as an existing product on 6 October 2004. Notification expired on 18 April 2007, with no application for renewal. Therefore, the authorisation is no longer valid. Regulatory approvals for MON80100 (*goxv247* plus *cry1Ab* plus *cp4 epsps*) were granted in the USA for the environment in 1995 and as food and/or feed in 1996. Environmental regulatory approvals were granted for GA21 maize in the USA in 1997, Argentina, Canada and Japan in 1998, Brazil in 2008 and the Philippines in 2009.

1.2.4 dicamba tolerance

Introduces tolerance to dicamba

Monsanto has announced that it has soybeans in development that are tolerant of the herbicide dicamba. There are no details of the nature of the gene. The company announced in July 2010 that it had completed regulatory submission to the US Department of Agriculture for dicamba-tolerant soybeans. Monsanto expects to complete regulatory submission to the US Food and Drug Administration and key global markets shortly.

In the first season of South American testing and the second overall season of testing, research indicates that dicamba-tolerant soybeans have demonstrated tolerance at both pre-emergent and post-emergent application timing, indicating that the trait is conveying the desired level of herbicide tolerance in the Phase II commercial-track events. Dicamba tolerance is the third-generation weed control trait that will be stacked with 'Roundup RReady2Yield' and the high-yielding 'Genuity Roundup Ready 2 Yield' soybean trait to give farmers the most advanced in-seed weed-control system available.

1.2.5 *2,4-D monooxygenase* gene

Introduces tolerance to 2,4-D

NOMENCLATURE: Approved name: *2,4-D monooxygenase* gene.

Other names: *tfdA* gene.

Promoter(s): Cauliflower mosaic virus (CaMV) 35S.

SOURCE: The gene for the 2,4-D monooxygenase enzyme is encoded on a large plasmid found in the *Alcaligenes eutrophus* bacterium, and was isolated as a small fragment of DNA.

PRODUCTION: The hybrid *tfdA* plus CaMV 35S gene was introduced into *Agrobacterium tumefaciens* and used to infect cotton (cultivar 'Coker'). 'Coker' cotton is a poor agronomic variety but has very good tissue culture characteristics, making it ideal for transformation work and it proved to be slightly easier to transform than Sikora cotton. The 'Coker' cultivar was adopted as a model substitute for commercial varieties, although useful genes introduced into 'Coker' varieties can still be moved into more commercial cultivars by traditional breeding methods. The first four fertile 'Coker' plants were analysed further, by genetic and molecular techniques, after transfer to the glasshouse and collection of progeny seeds. Two of the lines contained single insertions of the novel DNA, while the other two contained two insertions of the gene that segregated independently in the progeny.

TARGET CROPS: Cotton.

BIOLOGICAL ACTIVITY: The first step in the complex pathway of 2,4-D degradation is sufficient to detoxify 2,4-D in plants. This step is catalysed by the enzyme 2,4-D monooxygenase, which cleaves an acetate group off 2,4-D to produce dichlorophenol (DCP). DCP is at least 100 times less toxic to plants than 2,4-D.
Biology: When sprayed with 2,4-D, the GM-plants were tolerant of 2,4-D at 600 ppm, especially when in the pure homozygous form. The two heterozygous lines showed more extensive damage at 600 and 900 ppm 2,4-D. At 900 ppm, only slight damage occurred to the shoot tips of the best-performing line, and it is the homozygous progeny of this plant that were field-tested.

1.2 Herbicide tolerance

COMMERCIALISATION: Formulation: Elite Australian cotton varieties are being developed through traditional breeding techniques.

APPLICATION: Cotton is very sensitive to drift from 2,4-D applied to maize and cereals. These varieties are being evaluated for their potential in preventing drift damage.

COMPATIBILITY: GM-cotton can be treated with all the crop protection chemicals that are applied to a traditional crop.

MAMMALIAN TOXICITY: Cotton transformed with the *tfdA* gene is considered to be substantially identical to conventional varieties and is not considered to pose any mammalian risks. There have been no reports of allergic or other adverse toxicological effects from research workers, breeders or field staff.

ENVIRONMENTAL IMPACT AND NON-TARGET TOXICITY: As the transgenic crop is considered not to be substantially different from conventionally grown crops, it is not expected that it will show any adverse effects on non-target organisms or on the environment.

1.2.6 *gat4601*

Introduces tolerance to glyphosate

NOMENCLATURE: Approved name: *gat* gene.

Other names: *gat4621* gene; *glyphosate acetyltransferase* gene.

Development code(s): Soybean – DP-356Ø43-5 (DP356043); DP-356Ø43-5 (Event DP356043). Maize (corn) – DP-Ø9814Ø-6 (Event 98140); DP-Ø9814Ø-6 (Event 98140).

Promoter(s): In DP356043 – *gm-hra* S-adenosyl-L-methionine synthetase (SAMS); *gat4601* SCP1 – constitutive synthetic core promoter TMV omega 5′-UTR; 98140 – *gm-hra – als* (*Zea mays*) CaMV 35S enhancer element; *gat4621* ZmUbiInt (*Zea mays* polyubiquitin gene promoter and first intron). CaMV 35S enhancer element.

Trait introduction: DP356043 was produced by microprojectile-mediated bombardment of embryogenic somatic cell cultures, derived from explants from small, immature soybean seeds of the cultivar, 'Jack'.

SOURCE: The *gat4621* gene is based on the sequences of three *gat* genes from the common soil bacterium *Bacillus licheniformis* (Weigmann) Chester. The *gm-hra* gene cassette contains the coding sequence of the *gm-hra* gene, which encodes the GM-HRA protein. It is a modified version of the endogenous soybean acetolactate synthase gene (*gm-als*), which encodes the GM-ALS I protein.

TARGET PESTS: A wide variety of weeds.

TARGET CROPS: Soybean

BIOLOGICAL ACTIVITY: Maize line 98140 was genetically engineered to express the GAT4621 (glyphosate acetyltransferase) protein. The GAT4621 protein, encoded by the *gat4621* gene, confers tolerance to glyphosate-containing herbicides by acetylating glyphosate and thereby rendering it non-phytotoxic.

COMMERCIALISATION: Trade names: 'Optimum GAT Soybeans' (*gm-hra* plus *gat4601*) (Pioneer) and (DuPont), 'Optimum GAT Corn' (*gm-hra* plus *gat4601*) (Pioneer) and (DuPont).

PRODUCT SPECIFICATIONS: Soybean event 356043 has been field-tested in the USA, Canada, and Japan. During the field tests, several phenotypic characteristics were evaluated, including early population, seedling vigour, days to maturity, plant height, lodging, pod shattering, final population, and yield. The statistical analysis of these observations showed no biologically meaningful differences between soybean 356043 and the control plant, and supports a conclusion of phenotypic equivalence to currently commercialised soybean lines. The seed dormancy and germination of soybean event 356043 were compared with the control, and no significant differences were detected in percentage germinated seed, percentage dead seed, and percentage viable firm, swollen seed and viable hard seed. Viable hard seed was not observed in seed germination and seed viability tests of soybean event 356043. Data on the susceptibility to a range of insect pests and diseases, and response to abiotic stressors, were also recorded. No increase or decrease outside of the reference range was observed in any insect or abiotic stressor in soybean event 356043. Ecological observations (plant interactions with insects and diseases) were recorded for all USDA-APHIS-permitted field trials of 98140 maize during the 2005 and 2006 growing seasons. Plant breeders and field staff familiar with plant pathology and entomology observed event 98140 and control lines at least every four weeks for insect and disease pressure and recorded the severity of any stressor seen. In every case, the severity of insect or disease stress on event 98140 was not qualitatively different from various control lines growing at the same location. These results support the conclusion that the ecological interactions for event 98140 were comparable to control maize lines with similar genetics or to conventional maize lines.

COMPATIBILITY: Compatible with a wide variety of crop protection chemicals used in crop cultivation.

MAMMALIAN TOXICITY: Bioinformatics studies with the GAT4601 and GM-HRA proteins confirmed the absence of any biologically significant amino acid sequence similarity to known protein toxins or allergens, and digestibility studies demonstrated that both proteins would be rapidly degraded following ingestion, similarly to other dietary proteins. Acute oral toxicity studies in mice with both proteins also confirmed the absence of toxicity. Taken together, the evidence indicated that neither protein is toxic or likely to be allergenic in humans. The metabolite residues generated by glyphosate-treated soybean 356043 plants are considered less toxic than glyphosate, which itself is considered to be of very low potential toxicity in animals. Hence, there is no increase in overall toxicity arising from the presence of glyphosate residues on soybean 356043, and the current acceptable daily intake (ADI) for glyphosate is considered to be protective of public health and safety. The GAT4621 and GM-HRA proteins were evaluated for potential allergenicity and toxicity using a weight of evidence approach that examined: amino acid sequence similarities with known allergens and toxins, digestibility in model pepsin (simulated gastric fluid, SGF) and pancreatin (simulated intestinal fluid) systems, post-translational modification, and characteristics of the donor organism. Neither protein displayed significant amino acid sequence similarity with known allergens or toxins, both proteins were rapidly and completely digested in SGF within 30 seconds, and neither protein was glycosylated.

ENVIRONMENTAL IMPACT AND NON-TARGET TOXICITY: Considering that GAT4601 and GM-HRA proteins lack the properties of known toxins or allergens, there should be no significant impacts to the environment or to non-target organisms, including threatened and endangered species, from the introduction of 356043 soybean.

DATE(S) OF REGISTRATION: Environmental regulatory approvals for DP356043 were granted in the USA in 2008 and in Canada and Japan in 2009; food and/or feed uses were approved in the USA in 2007 and in Mexico in 2008; food uses were granted in Canada, Japan, Korea and the Philippines in 2009; and feed uses were approved in Canada, Japan, Taiwan and the Philippines in 2009 and in Australia in 2010. Environmental regulatory approvals for 'Optimum GAT Corn' were granted in the USA and Canada in 2009; food and/or feed approvals were granted in the USA in 2008; food use was allowed in Canada in 2009; and feed use in Canada in 2009 and Korea in 2010. Environmental regulatory approvals for DP356043 were allowed in the USA in 2008 and in Canada and Japan in 2009; food and/or feed approvals were granted in the USA in 2007 and in Mexico in 2008; food uses were approved in Canada, Japan, the Philippines and Taiwan in 2009 and in Australia in 2010; feed uses were approved in Canada, Japan, Korea and the Philippines in 2009. Environment regulatory approvals were allowed for DP98140 in the USA and Canada in 2009; food and/or feed uses were approved in the USA in 2008; food and feed uses were approved in Canada in 2009; and feed use was approved in Korea in 2010.

1.2.7 *gm-hra*

Introduces tolerance to als-inhibiting herbicides

NOMENCLATURE: Approved name: *gm-hra* gene.

Development code(s): Soybean – DP-356Ø43-5 (DP356043); DP-356Ø43-5 (Event DP356043). Maize (corn) – DP-Ø9814Ø-6 (Event 98140); DP-Ø9814Ø-6 (Event 98140).

Promoter(s): DP356043 – *gm-hra* – S-adenosyl-L-methionine synthetase (SAMS); *gat4601* SCP1 – constitutive synthetic core promoter TMV omega 5′-UTR; 98140 – *gm-hra* – *als* (*Zea mays*) CaMV 35S enhancer element; *gat4621* – ZmUbiInt (*Zea mays* polyubiquitin gene promoter and first intron). CaMV 35S enhancer element.

Trait introduction: DP356043 was produced by microprojectile-mediated bombardment of embryogenic somatic cell cultures, derived from explants from small, immature soybean seeds of the cultivar, 'Jack'.

SOURCE: The *gm-hra* gene cassette contains the coding sequence of the *gm-hra* gene, which encodes the GM-HRA protein. It is a modified version of the endogenous soybean acetolactate synthase gene (*gm-als*), which encodes the GM-ALS I protein. The *gat4621* gene is based on the sequences of three *gat* genes from the common soil bacterium *Bacillus licheniformis* (Weigmann) Chester.

PRODUCTION: The herbicide-tolerant *gm-hra* gene was made by isolating the herbicide-sensitive soybean *gm-als* gene and introducing two specific amino acid changes known to confer herbicide tolerance to tobacco ALS.

TARGET PESTS: A wide variety of weeds.

TARGET CROPS: Soybean and maize (corn).

BIOLOGICAL ACTIVITY: The herbicide-tolerant *gm-hra* gene is not inhibited by imidazolinone herbicides.

COMMERCIALISATION: Trade names: 'Optimum GAT Soybeans' (*gm-hra* plus *gat4601*) (Pioneer) and (DuPont), 'Optimum GAT Corn' (*gm-hra* plus *gat4601*) (Pioneer) and (DuPont).

PRODUCT SPECIFICATIONS: Soybean event 356043 has been field-tested in the USA, Canada, and Japan. During the field tests, several phenotypic characteristics were evaluated, including early population, seedling vigour, days to maturity, plant height, lodging, pod shattering, final population, and yield. The statistical analysis of these observations showed no biologically meaningful differences between soybean 356043 and the control plant, and supports a conclusion of phenotypic equivalence to currently commercialised soybean lines. The seed dormancy and germination of soybean event 356043 were compared with the control, and no significant differences were detected in percentage germinated seed, percentage dead seed, and percentage viable firm, swollen seed and viable hard seed. Viable hard seed was not observed in seed germination and seed viability tests of soybean event 356043. Data on the susceptibility to a range of insect pests and diseases, and response to abiotic stressors, were also recorded. No increase or decrease outside of the reference range was observed in any insect or abiotic stressor in soybean event 356043. Ecological observations (plant interactions with insects and diseases) were recorded for all USDA-APHIS-permitted field trials of 98140 maize during the 2005 and 2006 growing seasons. Plant breeders and field staff familiar with plant pathology and entomology observed event 98140 and control lines at least every four weeks for insect and disease pressure and recorded the severity of any stressor seen. In every case, the severity of insect or disease stress on event 98140 was not qualitatively different from various control lines growing at the same location. These results support the conclusion that the ecological interactions for event 98140 were comparable to control maize lines with similar genetics or to conventional maize lines.

COMPATIBILITY: Compatible with a wide variety of crop protection chemicals used in crop cultivation.

MAMMALIAN TOXICITY: Bioinformatics studies with the GAT4601 and GM-HRA proteins confirmed the absence of any biologically significant amino acid sequence similarity to known protein toxins or allergens, and digestibility studies demonstrated that both proteins would be rapidly degraded following ingestion, similarly to other dietary proteins. Acute oral toxicity studies in mice with both proteins also confirmed the absence of toxicity. Taken together, the evidence indicated that neither protein is toxic or likely to be allergenic in humans. The metabolite residues generated by glyphosate-treated soybean 356043 plants are considered less toxic than glyphosate, which itself is considered of very low potential toxicity in animals. Hence, there is no increase in overall toxicity arising from the presence of glyphosate residues on soybean 356043, and the current acceptable daily intake (ADI) for glyphosate is considered to be protective of public health and safety. The GAT4621 and GM-HRA proteins were evaluated for potential allergenicity and toxicity using a weight of evidence approach that examined: amino acid sequence similarities with known allergens and toxins, digestibility in model pepsin (simulated gastric fluid, SGF) and pancreatin (simulated intestinal fluid) systems, post-translational modification and characteristics of the donor organism. Neither protein displayed significant amino acid sequence similarity with known allergens or toxins, both proteins were rapidly and completely digested in SGF within 30 seconds, and neither protein is glycosylated.

ENVIRONMENTAL IMPACT AND NON-TARGET TOXICITY: Considering that GAT4601 and GM-HRA proteins lack the properties of known toxins or allergens, there should be no significant impacts to the environment or to non-target organisms, including threatened and endangered species, from the introduction of 356043 soybean.

DATE(S) OF REGISTRATION: Environmental regulatory approvals for DP356043 were granted in the USA in 2008 and in Canada and Japan in 2009; food and/or feed uses were approved in the USA in 2007 and in Mexico in 2008; feed uses were granted in Canada, Japan, Korea and the Philippines in 2009; and feed uses were approved in Canada, Japan, Taiwan and the Philippines in 2009 and in Australia in 2010. Environmental regulatory approvals for 'Optimum GAT Corn' were granted in the USA and Canada in 2009; food and/or feed approvals were granted in the USA in 2008; food use was allowed in Canada in 2009; and feed use in Canada in 2009 and Korea in 2010. Environmental regulatory approvals for DP356043 were allowed in the USA in 2008 and in Canada and Japan in 2009; food and/or feed approvals were granted in the USA in 2007 and in Mexico in 2008; food uses were approved in Canada, Japan, the Philippines and Taiwan in 2009 and in Australia in 2010; feed uses were approved in Canada, Japan, Korea and the Philippines in 2009. Environment regulatory approvals were allowed for DP98140 in the USA and Canada in 2009; food and/or feed uses were approved in the USA in 2008; food and feed uses were approved in Canada in 2009; and feed use was approved in Korea in 2010.

1.2.8 *gox247*

Introduces tolerance to glyphosate

NOMENCLATURE: Approved name: *gox* gene.

Other names: *goxv247*.

Development code(s): Sugar beet – GTSB77.

Promoter(s): canola GT200, GT73, RT73, ZSR500/502 (*cp4 epsps* and *goxv247*) – chloroplast transit peptide from *Arabidopsis thaliana*.

SOURCE: Sugar beet line GTSB77 expresses a herbicide-tolerant form of the enzyme 5-enolpyruvylshikimate-3-phosphate synthase (EPSPS) derived from the common soil bacterium, *Agrobacterium tumefaciens* Conn. strain CP4. The plasmid used during the transformation of this sugar beet line also contained a second gene that encoded the enzyme glyphosate oxidase (GOX) from the bacterium *Ochrobactrum anthropi*. This enzyme was intended to accelerate the normal degradation of glyphosate into aminomethylphosphonic acid (AMPA) and glyoxylate in plant cells. Upon insertion into the plant genome, the GOX-encoding gene was truncated and fused to genomic sugar beet DNA sequences, resulting in a chimeric sequence that did not express a functional protein.

TARGET PESTS: A wide variety of weeds.

TARGET CROPS: Sugar beet, canola, maize (corn) and cotton.

BIOLOGICAL ACTIVITY: Mode of action: The enzyme glyphosate oxidase (GOX) isolated from the bacterium *O. anthropi* accelerates the normal degradation of glyphosate into aminomethylphosphonic acid (AMPA) and glyoxylate in plant cells.

COMMERCIALISATION: Trade names: 'InVigor' (*cp4 epsps* plus *goxv247*) (Monsanto) and (Syngenta), 'Roundup Ready Canola' (*cp4 epsps* plus *goxv247*) (Monsanto), 'Westar Roundup Ready Canola' (*cp4 epsps* plus *goxv247*) (Monsanto), 'Hysyn 101 RR Roundup Ready Canola' (*cp4 epsps* plus *goxv247*) (Monsanto), 'Roundup Ready Corn 2' (*cp4 epsps* plus *goxv247*) (MON832) (Monsanto).

COMPATIBILITY: The transgenic sugar beet line GTSB77 was field tested in the USA (1996–98). GTSB77 was evaluated extensively, and agronomic data showed that there were no significant changes in the number of seeds produced, germination characteristics, final stand, overwintering capability, or pathogen susceptibility.

MAMMALIAN TOXICITY: Crops containing the *gox* gene are considered to be substantially identical to conventional varieties and are not considered to pose any mammalian risks. There have been no reports of allergic or other adverse toxicological effects from research workers, breeders or field staff.

ENVIRONMENTAL IMPACT AND NON-TARGET TOXICITY: It has been concluded that the *gox* gene will not result in any deleterious effects or significant impacts on non-target organisms, including threatened and endangered species or beneficial organisms.

DATE(S) OF REGISTRATION: Regulatory approvals for MON80100 (*goxv247* plus *cry1Ab* plus *cp4 epsps*) were granted in the USA for the environment in 1995 and as food and/or feed in 1996. Regulatory approvals for MON802 (*goxv247* plus *cry1Ab* plus *cp4 epsps*) were granted for the environment in Canada, Japan and the USA in 1997; for food and/or feed in the USA in 1996; and for food and use as feed in Canada in 1997. Regulatory approvals for MON809 (*goxv247* plus *cry1Ab* plus *cp4 epsps*) for the environment were granted in Canada and the USA in 1996 and in Japan in 1997; for food and/or feed in the USA in 1996; in Canada for food in 1996; and for use as feed in Canada in 1996 and in Japan in 1998. Environment regulatory approvals were granted for 'InVigor Sugar Beet' in the USA in 1998; food and/or feed approvals were granted in the USA in 1998 and Australia in 2002; food uses were allowed in Japan in 2003; and food and feed uses were approved in the Philippines in 2004.

1.2.9 *IMI tolerance* gene

Introduces tolerance to imidazolinone herbicides

NOMENCLATURE: Approved name: *IMI tolerance* gene.

Other names: *IMI* gene; *imidazolinone tolerance* gene; *sulfonylurea tolerance* gene; *STS* gene; *als* gene; *als 1* gene.

Development code(s): Canola – NS738; NS1471; NS1473.
Carnation – 4, 11, 15, 16, 66, 959A, 988A, 1226A, 1351A, 1363A and 1400A.
Cotton – DD-Ø1951A-7 (19-51A). Sunflower – X81359. Lentil – RH44. Linseed – CDC-FLØØ1-2 (FP967). Rice – CL121, CL141, CFX51, IMINTA-1, IMINTA-4, PWC16. Wheat – AP205CL, AP602CL, BW255-2, BW238-3, BW7, SWP965001 and Teal 11A. Maize (corn) – 3751IR, DP-98140-6, EXP1910IT, IT.

Promoter(s): Not relevant in canola, cotton, linseed, sunflower, maize (corn) and wheat as the trait was selected rather than introduced. In carnation – CaMV 35S.

Trait introduction: Canola – selection of somaclonal variants from microspore cultures following chemical mutagenesis. Carnation – the transgenic carnation lines 4, 11, 15, 16, 959A, 988A, 1226A, 1351A, 1363A and 1400A were produced by *Agrobacterium*-mediated transformation of carnation plants encoding the *surB* (ALS encoding) gene from a chlorsulfuron tolerant line of tobacco. Cotton 19-51A trait introduction was by *Agrobacterium tumefaciens*-mediated plant transformation. Sunflower – the imidazolinone-tolerance trait in X81359 was discovered in a wild (weedy) population of sunflower (*Helianthus annuus* L.). The natural mutation conferring the imidazolinone tolerance was detected in a field where soybeans had been cultivated and treated with imazethapyr for several seasons. The mutant sunflower population expressed a mutation in the *als* gene, conferring tolerance to applications of imazethapyr. Conventional plant-breeding techniques were used to introduce the herbicide-tolerance trait into sunflower germplasm, including extensive backcrossing into 'NuSun' sunflower varieties. Lentil – RH44 lentil was isolated from a population derived from the seed of several lentil cultivars treated with ethylmethane sulfonate (EMS), a substance known to induce point mutations within the genome of organisms. The selection of herbicide-tolerant plants was made from whole plants treated with an imidazolinone herbicide. One imidazolinone-tolerant plant was selected from this population. Conventional breeding and seed increase techniques were then used to develop the imidazolinone-tolerant line RH44. Rice – CL121, CL141 and CFX51 chemically induced seed mutagenesis using ethylmethane sulfonate and traditional cross-breeding; IMINTA-1 and IMINTA-4 were developed through chemically induced mutagenesis of rice variety IRGA 417 with sodium azide; PWC16 was isolated from a population derived from seed of the variety 'Cypress', which had been treated with ethylmethane sulfonate. Wheat – the wheat line AP205CL was isolated from a population derived from seed of the variety 'Gunner', which had been treated with ethylmethane sulfonate and diethyl sulfate. The selection of herbicide-tolerant plants was made from whole plants treated with imazamox. The designation AP205CL was given to the imazamox-tolerant plant selected from the population. AP602CL was selected from a population of wheat derived by chemically induced mutagenesis of seed of the wheat variety 'Gunner'. One mutant tolerant to imidazoline was selected and designated AP602CL. The trait in BW255-2, BW238-3 and BW7 was developed using chemically induced seed mutagenesis and whole-plant selection procedures. SWP965001 was isolated was from a population derived by chemically induced mutagenesis of seed of the winter wheat cultivar 'Fidel' with sodium azide. Four imidazolinone-tolerant seedlings, FSI, FS2, FS3 and FS4, were selected in the presence of herbicide. SWP965001 is derived from the selection line FS2 by an initial cross to the spring wheat cultivar 'Grandin', followed by two backcrosses to 'Grandin'. The wheat line Teal 11A was isolated from a population derived from seed of the variety 'CDC Teal', which had been treated with ethylmethane sulfonate. The selection of herbicide-tolerant plants was made from whole plants treated with imazamox and imazethapyr. The designation Teal 11A was given to the mutant that had exhibited tolerance to both herbicides. Seed from this line was obtained from self-fertilisation. Somatic maize embryos grown on imidazolinone-enriched media were selected and, from these, the somaclonal variant cell line XA17 was isolated. This cell line was regenerated to a whole plant and crossed to the inbred line B73. The XA17/B73 line was backcrossed into each of two proprietary Pioneer inbreds, which were then crossed to produce the hybrid 3751IR. DP-98140-6 maize contains the *als* gene plus the *hra* gene. The imazethapyr-tolerant trait in line EXP1910IT was selected following chemical mutagenesis by exposing pollen to ethyl-methane sulfonate. Mutagenised pollen was then used to fertilise the parent line, UE95, and progeny plants were screened for tolerance to imazethapyr. Pioneer International licensed an imidazolinone-tolerant maize line from American Cyanamid (now BASF) that was used to develop new corn hybrids with herbicide tolerance. The mutation is known as XI-12 and the resulting hybrids as IT hybrids. The IT hybrids are very similar to imidazolinone-resistant (IR) hybrids in their mechanism of herbicide resistance. The method used for obtaining these imidazolinone-tolerant maize plants was *in vitro* selection.

SOURCE: In the 1980s, Cyanamid (now BASF) funded Molecular Genetics to select maize cell lines that were tolerant of imidazolinone herbicides, using *in vitro* selection techniques. These cell lines were grown on to produce mature maize plants that were shown to possess a modified acetolactate synthase gene coding for an enzyme that was not inhibited by the imidazolinones. These mature plants were used as one parent in a breeding programme undertaken by Pioneer International, that led to the launch of IMI-corn in the USA in 1995. The same technique has been used to produce other IMI-tolerant crops.

TARGET PESTS: All types of weed.

TARGET CROPS: Maize (corn), canola, cotton, rice, soybean, carnation, sunflower and wheat.

BIOLOGICAL ACTIVITY: Imidazolinone herbicides are active against the enzyme acetolactate synthase (ALS), also known as acetohydroxyacid synthase (AHAS) or acetolactate pyruvate-lyase. This enzyme catalyses the first step in the biosynthesis of the essential branched-chain amino acids isoleucine, leucine and valine. Herbicide-induced ALS inhibition results in a lethal decrease in protein synthesis. Unmodified canola is not tolerant to imidazolinones. The genome of *Brassica napus* contains an ALS multigene family comprising five genes, two of which are constitutively expressed, share extensive homology with ALS genes from *Arabidopsis thaliana* and *Nicotiana tabacum*, and are assumed to encode the primary ALS activities necessary for plant growth and development. Modifications of ALS genes in various plant species can result in herbicide-tolerant phenotypes and typically consist of one amino acid substitution, sufficient to alter the binding site for imidazolinones such that the herbicide no longer inactivates the ALS enzyme. Pioneer International Inc. selected two *B. napus* lines, P1 and P2, with altered ALS enzymes not inhibited by the herbicide. The modifications in P1 and P2 occur at different and unlinked loci, presumably representing different ALS genes. Lines NS1471 and NS1473 contain both mutations, while line NS738 contains the P2 mutation only. The novel ALS enzyme in line P2 has been isolated and sequenced: it differs from the native gene by one amino acid substitution. The novel imidazolinone resistance is constitutively expressed, and synthesis of the branched-chain amino acids isoleucine, leucine and valine is not affected by the ALS modifications in NS1471, NS1473 and NS738. These lines were developed from the *in vitro* culture of microspores, from the *B. napus* cultivar 'Topas', mutagenised with ethylnitrosourea and selected on imidazolinone containing culture medium. The regenerated haploid plantlets were recovered and treated with colchicine to induce chromosome duplication, resulting in doubled-haploid plants. The two imidazolinone-tolerant lines P1 and P2 were selected and crossed reciprocally then crossed with *B. napus* cultivars 'Topas' and 'Regent', and the progeny subjected to repeated cycles of selfing and further breeding. **Biology:** The plants are tolerant of the imidazolinone group of herbicides by virtue of an enzyme that is not inhibited by the herbicides. **Mode of action:** The selected plants contain an acetolactate synthase enzyme that is not inhibited by the imidazolinone herbicides and, as such, they can tolerate over-the-top applications of these herbicides. American Cyanamid (now BASF) held a major share of the soybean herbicide market with these compounds, but none was selective in maize. The introduction of these varieties allowed these herbicides to be used in maize crops. It has been shown that IMI-maize and sorghum varieties can be rendered tolerant to attack by *Striga* species, if the seed is treated with an imidazolinone herbicide. Such techniques are being developed in Africa by BASF. **Efficacy:** Tolerance is so complete that there have been reports of imazethapyr-tolerant maize occurring in conventional soybean crops. The University of Saskatchewan, Crop Development Centre is developing linseed FF967 for use in soils containing residues of triasulfuron and metsulfuron-methyl in areas of Canada where flax is usually cultivated.

COMMERCIALISATION: Formulation: Introduced by Pioneer and Garst in 1995 and now commercialised by over 60 seed companies in the USA. **Trade names:** 'Clearfield Corn' (*als*) (Agripon Seeds), (AgVenture Seed), (Asgrow Seed), (Becks Hybrid Seeds), (Callahan Seeds), (Cargill Hybrid Seeds), (Chemgro Seeds), (Countrymark Cooperative), (Croplan Genetics), (DeKalb), (Garst Seed), (Golden Harvest Seeds), (Growmark), (Hoegemeyer Hybrids), (Hoffman Seeds), (Interstate Seed), (Midwest Seed Genetics), (Mycogen), (NC+ Hybrids), (Pioneer International), (Sands of Iowa), (Scotts Quality Seeds), (Trisler Seed Farms) and (Wilson Seeds), 'Clearfield Corn plus Bt' (*als* plus *cry1A(c)*) (Asgrow Seed), (Becks Hybrid Seeds), (Garst Seed), (Mycogen) and (Syngenta Seeds), 'Clearfield 45A71 Canola' and 'Clearfield 45A72 Canola' (Pioneer) and (Advanta Pacific Seeds), 'Clearfield Wheat' (*als*), 'Clearfield Sugar Beet' (*als*), 'Clearfield Rice (*als*)', 'Clearfield Sunflower' (*als*) and 'Clearfield Sugar Cane' (*als*) (BASF), 'Carnation – 4, 11, 15, 16' (*surB* plus *dfr* plus *hfl*), 'Carnation 66' (*surB* plus *ACC*) and 'Carnation – 959A, 988A, 1226A, 1351A, 1363A and 1400A' (*surB* plus *dfr* plus *bp40*)' (Florigen), 'Cotton 19-51a' (*als*) (DuPont) and (Pioneer), 'CDC Triffid' (*als*) (University of Saskatchewan), 'Optimun GAT' (*als* plus *hra*) (DuPont) and (Pioneer), 'STS tolerant Soybeans (Agratech Seeds),

(Agripro Seeds), (AgVenture), (Asgrow Seeds), (Becks Hybrids), (Campbell Seeds), (Chemgro Seeds), (Countrymark Cooperative), (Croplan Genetics), (Dairyland Seed Co.), (DeKalb Genetics), (Deltapine Seed), (Garst Seed), (Golden Harvest Seeds), (Growmark), (Hoegemeyer Hybrids), (Hoffman Seeds), (Latham Seeds), (Merscham Seeds), (Midwest Seed Genetics), (Mycogen), (NC+ Hybrids), (Pioneer), (Rupp Seed), (Sands of Iowa), (Scotts Quality Seeds) and (Sieben Hybrids), 'STS tolerant Cotton' (Du Pont), 'STS tolerant Canola' (Cargill Hybrid Seeds).

PRODUCT SPECIFICATIONS: Selected imidazolinone-tolerance trait bred into elite maize varieties.

COMPATIBILITY: Compatible with all imidazolinone herbicides and herbicides generally used to control weeds in maize. Volunteer crops can be controlled by the use of herbicides that are used to remove conventional varieties, with the exception of imidazolinone herbicides.

MAMMALIAN TOXICITY: IMI-tolerant varieties have been shown to be substantially identical to related varieties and are not expected to show any adverse toxicological effects. There have been no reports of allergic or other adverse toxicological effects from research workers, breeders or growers. The risk of transferring genetic traits from transgenic carnation lines to species in unmanaged environments is insignificant. The transgenic carnation lines are intended for cultivation by growers, flower auctions, flower wholesalers, retailers and breeders. Plants are sold as flowers, cuttings or whole plants.

ENVIRONMENTAL IMPACT AND NON-TARGET TOXICITY: IMI-tolerant varieties are not expected to have any adverse effects on non-target organisms or on the environment.

DATE(S) OF REGISTRATION: 'Clearfield Canola' was registered in Canada for the environment and for food and feed uses in 1995. Carnation – lines 4, 11, 15, 16 were given environmental approval in Australia in 1995 and marketing approval in the EU in1997. 'Carnation 66'was given environmental approval in Australia in 1995 and environmental and marketing approval in the EU in 1998. 'Carnations 959A, 988A, 1226A, 1351A, 1363A and 1400A' were given environmental and marketing approval in the EU in 1998 and environmental approval in Colombia in 2000. Line 19-51a was approved in the USA for the environment and for food and/or feed use in 1996. 'Clearfield Sunflower' was approved for food use in 2003 and for food and the environment in 2005 in Canada. The lentil line RH44 gained regulatory approval for the environment, food and feed uses in Canada in 2004. 'Clearfield Rice (CL121, CL141 and CFX51)' were approved for food and feed uses in Canada in 2002. 'Clearfield Rice IMINTA 1 and IMINTA 4' were approved for food and feed uses in Canada in 2006. 'Clearfield Rice PWC16' was approved for use as feed in 2002 and as food in 2003 in Canada. AP205CL ('Clearfield Spring Wheat') was approved for food use in 2003 and for the environment and feed use in Canada in 2004. AP602CL 'Clearfield

Spring Wheat' was approved for the environment, food and feed uses in Canada in 2003. BW255-2 and BW238-3 'Clearfield Bread Wheat' was approved for the environment and for food and feed uses in Canada in 2006. BW7 'Clearfield Bread Wheat' was approved for the environment and for food and feed uses in Canada in 2007. 'Clearfield Wheat SWP965001' was approved in Canada for the environment in 1998 and for food and feed uses in 1999. 'Clearfield Spring Wheat' (Teal 11A) received approval for the environment and for food and feed uses in Canada in 2004. 3751IR maize was approved in Canada for food use in 1994 and for the environment and feed use in 1996. 'Optimum GAT' was approved in the USA in 2009. The EXP1910IT line of corn was approved in Canada for the environment and for use as feed in 1996 and for use as food in 2007. IT maize hybrids were approved for food use in Canada in 1998.

1.2.10 *phosphinothricin acetyl transferase* gene

Introduces tolerance to glufosinate-ammonium

NOMENCLATURE: Approved name: *phosphinothricin acetyl transferase* gene.

Other names: *Bilanafos acetyl reductase* gene; 'Liberty Link' gene; *pat* gene; *bar* gene.

Development code(s): Sugar beet – ACS-BVØØ1-3 (T120-7). Canola – HCN10, ACS-BNØØ4-7 × ACS-BNØØ1-4 (MS1, RF1 → PGS1); ACS-BNØØ4-7 × ACS-BNØØ2-5 (MS1, RF2 → PGS2), ACS-BNØØ5-8 × ACS-BNØØ3-6 (MS8 × RF3), PHY14, PHY35, PHY36, ACS-BNØØ8-2 (T45 (HCN28)) and HCR-1. Chicory – RM3-3, RM3-4 and RM3-6. Soybean – ACS-GMØØ5-3 (A2704-12, A2704-21, A5547-35), ACS-GMØØ6-4 (A5547-127) and ACS-GMØØ3-1 (GU262). Cotton – ACS-GHØØ1-3 (LLCotton25), DAS-24236-5 (281-24-236), DAS-21Ø23-5 (3006-210-23), DAS-21Ø23-5 × DAS-24236-5, DAS-21Ø23-5 × DAS-24236-5 × MON-Ø1445-2, DAS-24236-5, DAS-21Ø23-5, MON-88913-8 (DAS-21Ø23-5 × DAS-24236-5 × MON88913), ACS-GHØØ1-3, MON-15985-7 (LLCotton25 × MON15985), Maize (corn) – 676, 678, 680, B16 (DLL25), BT11 (X4334CBR, X4734CBR), BT11 × GA21, BT11 × MIR162, BT11 × MIR162 × MIR604, BT11 × MIR604, BT11 × MIR604 × GA21, DAS-06275-8, DAS-59122-7, DAS-59122-7 × NK603, DAS-59122-7 × TC1507 × NK603, DBT418, MON89034 × TC1507 × MON88017 × DAS-59122-7, MS3, MS6, NK603 × T25, T14, T25, T25 × MON810, TC1507, TC1507 × DAS-59122-7, TC1507 × NK603.

Promoter(s): Cauliflower mosaic virus (CaMV) 35S. MS1, RF1 → PGS1; canola – *pat* PSsuAra from *Arabidopsis thaliana* chloroplast transit peptide; *barnase* – pTa 29 pollen-specific promoter from *Nicotiana tabacum*; *barstar* – anther-specific promoter.

Trait introduction: Event T120-7 sugar beet was produced by *Agrobacterium*-mediated transformation of calli from the parent line R01 with plasmid vector pOCA18/Ac. The transfer-DNA (T-DNA) portion of the tumour-inducing (Ti) bacterial plasmid was engineered to contain a modified form, the *pat*-encoding gene. HCN92 canola was derived from traditional breeding crosses between non-genetically modified canola and a line resulting from transformation event ('Topas 19/2'). The transformed line was initially crossed with line ACSN3, followed by a second cross with cultivar 'AC Excel'. The seed from the final cross was advanced and single seed from F3 plants were bulked to form HCN92. Canola line HCN28 was produced by *Agrobacterium*-mediated transformation of canola cultivar 'AC Excel'. The *pat* gene used in transformation event T45 is a synthetic version of the gene isolated from *Streptomyces viridochromogenes*. The original transformant (T45) was backcrossed twice with canola line 'AC Excel'. HCN28 was developed from the resultant BC2 using the single seed descent method. The *Brassica rapa* canola hybrid, HCR-1, was derived from an interspecific cross with the *B. napus* transformation event T45 that expresses the *pat* gene. The soybean lines A2704-12, A2704-21, and A5547-35 were produced via biolistic transformation of soybean lines with a pUC19-based plasmid containing a modified form of the *pat* gene under the control of promoter and termination sequences derived from the 35S transcript from cauliflower mosaic virus (CaMV). The soybean line A5547-127 was produced via biolistic transformation of soybean with a pUC19-based plasmid containing a modified form the *pat* gene under the control of promoter and termination sequences derived from the 35S transcript from cauliflower mosaic virus (CaMV). The cotton lines 281-24-236 and 3006-210-23 were produced by *Agrobacterium*-mediated transformation of plant cells from the cotton variety 'Germain's Acala GC510'. The stacked cotton line 'WideStrike' × MON1445 expresses four novel proteins: the delta-endotoxins Cry1F and Cry1Ac, which confer resistance to the lepidopteran pests of cotton, such as the cotton bollworm, pink bollworm and tobacco budworm; the CP4 EPSPS protein, which confers tolerance to the herbicide glyphosate; and the PAT protein, which confers tolerance to glufosinate-ammonium. The insecticidal proteins Cry1F and Cry1Ac are produced by the *cry1F* and *cry1Ac* genes, respectively, both of which are from 'WideStrike'. The cotton line 'WideStrike' was produced by crossing line 281-24-236 with line 3006-210-23. The PAT protein in 'WideStrike' is produced by the *pat* gene: this expressed novel protein was intended solely for use as a selectable marker during plant transformation. The CP4 EPSPS protein is produced by the *cp4 epsps* gene from MON1445. 'WideStrike' × MON88913 was produced by crossing 'WideStrike' insect-resistant cotton with the herbicide-tolerant cotton line MON88913. This stacked cotton line is a product of traditional plant breeding. LLCotton25 was developed by introducing the bilanafos resistance (*pat*) gene into the cotton variety 'Coker312', using *Agrobacterium*-mediated plant transformation. LLCotton25 × MON15985 cotton is the product of the cross-breeding of the herbicide-tolerant cotton line LLCotton25 with the insect-resistant cotton line 15985 (MON-15985-7). This stacked cotton line is a product of traditional plant breeding.

SOURCE: The *bilanafos acetyl reductase* (*bar*) gene was isolated by De Block *et al.* in 1987 and was shown to be identical to the *phosphinothricin acetyl transferase* (*pat*) gene isolated from the source organism of bilanafos, *S. viridochromogenes*, in 1988 by Strauch *et al.* The gene is now synthesised, rather than derived from the host bacterium, but is identical in its action. It is often used in plant transformation to act as a selectable marker, with tolerance to glufosinate in the field a bonus effect. Plant Genetic Systems (now Bayer CropSciences) have used the gene to develop a range of plant products that are tolerant to its herbicide. The promoter was isolated from the cauliflower mosaic virus and has been found to be very effective at enhancing transcription levels of foreign genes in plants. The transgenic line MS1 (B91-4) was produced by genetically engineering plants to be male sterile and tolerant to the herbicide glufosinate-ammonium (as a selectable marker). The parental line MS1 contains the *barnase* gene for male sterility, isolated from *Bacillus amyloliquefaciens*. The *barnase* gene encodes for a ribonuclease enzyme (RNAse) expressed only in the tapetum cells of the pollen sac during anther development. The RNAse affects RNA production, disrupting normal cell functioning and arresting early anther development, thus leading to male sterility. The transgenic line RF1 (B93-101) was produced by genetically engineering plants to restore fertility in the hybrid line and to be tolerant to the herbicide glufosinate-ammonium (as a selectable marker). Transgenic RF1 plants contain the *barstar* gene isolated from *B. amyloliquefaciens*. The *barstar* gene codes for a ribonuclease inhibitor (barstar enzyme) expressed only in the tapetum cells of the pollen sac during anther development. The ribonuclease inhibitor (barstar enzyme) specifically inhibits barnase RNAse expressed by the MS1 line. Together, the RNAse and the ribonuclease inhibitor form a very stable one-to-one complex, in which the RNAse is inactivated. As a result, when pollen from the restorer line RF1 is crossed to the male sterile line MS1, the resultant progeny express the RNAse inhibitor in the tapetum cells of the anthers, allowing hybrid plants to develop normal anthers and restore fertility. Both transgenic canola lines MS1 and RF1 contain the *pat* gene.

PRODUCTION: The MS1 and RF1 and MS1 and RF2 canola lines were developed using genetic engineering techniques to provide a pollination control system for production of hybrid oilseed rape (MS1 × RF1). The novel hybridisation system involves the use of two parental lines, a male sterile line MS1 and a fertility restorer line RF1 (RF2). The transgenic MS1 plants do not produce viable pollen grains and cannot self-pollinate. In order to completely restore fertility in the hybrid progeny, line MS1 must be pollinated by a modified plant containing a fertility restorer gene, such as line RF1 (RF2). The resultant F1 hybrid seed, derived from a cross between MS1 × RF1 (RF2), produces hybrid plants that produce pollen and are completely fertile. MON89034 × TC1507 × MON88017 × DAS-59122-7 (*pat* plus *cp4 epsps* plus *cry1A.105* plus *cry2Ab* plus *cry3Bb1* plus *cry34Ab1* plus *cry35Ab1* plus *cry1Fa2*). BT11 × MIR604 × GA21 (*pat* plus *cp4 epsps* plus *cry3A* plus *cry1Ab*)

TARGET PESTS: A wide variety of weeds with some containing *Bt*-genes for insect pest control.

TARGET CROPS: This gene codes for an enzyme that catalyses a highly specific acetylation of the herbicide glufosinate, yielding the herbicidally inactive metabolite *N*-acetyl-L-glufosinate. The gene has been used to transform a variety of different crops.

BIOLOGICAL ACTIVITY: Biology: The transformed crops are tolerant of applied glufosinate herbicide because they are able to acetylate it very rapidly into a non-phytotoxic metabolite. **Mode of action:** The bacterium that produces the related natural herbicide, bilanafos (or phosphinothricin) (see *The Pesticide Manual* Fifteenth Edition, entry 444), produces an enzyme that acetylates the herbicide as a defence mechanism from its effect on the enzyme glutamine synthetase. This enzyme was isolated from *S. viridochromogenes* that produces it and was used to transform elite varieties of several crops. These transformed crops are able to acetylate the herbicide when it is sprayed on them and escape its effects. **Efficacy:** The gene expressed in transgenic crops renders the crop tolerant of over-the-top application of the herbicide glufosinate-ammonium, as it is acetylated very rapidly before any deleterious effects on the enzyme glutamine synthetase can occur.
Key reference(s): 1) M De Block, J Bottermann, M Vandewiele, T Dockx, C Thoen, V Gossele, N R Movva, C Thompson, M Van Montagu & J Leemans. 1987. Engineering herbicide resistance into plants by expression of a detoxifying enzyme, *EMBO Journal*, **6**, 2513–8. 2) E Strauch, W Arnold, R Alijah, W Wohlleben, A Pühler, G Donn, E Uhlmann, F Hein & F Wengenmayer. 1988. *Chemical Abstracts*, **110**, 34815z. 3) E Rasche & M Gadsby. 1997. Glufosinate-ammonium tolerant crops – international developments and experiences, *Proc. Brighton Crop Protection Conference – Weeds*, **3**, 941–6.

COMMERCIALISATION: Formulation: 'Liberty Link' canola was the first transgenic crop to be introduced commercially when it was approved by Canadian authorities in 1995. Introductions of glufosinate-tolerant maize (corn) in the USA followed in 1997 and soybean and sugar beet crops will be commercialised in the USA shortly. European registrations for transgenic oilseed rape, maize and sugar beet have been delayed and, in some cases, abandoned ('Chardon' maize) in Europe because of public resistance to GM-crop registrations. **Trade names:** 'Liberty Link Canola' (Bayer) (originally Plant Genetic Systems), 'Liberty Link Cotton' (Bayer), 'Liberty Link Sugar Beet' (Bayer), 'Liberty Link Soybeans' (Bayer), 'InVigor' (Bayer), 'Chardon LL' (forage maize) (Bayer), 'T25' (maize) (Bayer), 'Liberty Link Field Corn' (AgVenture), (Asgrow), (Callahan Seeds), (Campbell Seeds), (Cargill Hybrid Seeds), (Chemgro Seeds), (Croplan Genetics), (DeKalb), (Garst Seed), (Golden Harvest Seeds), (Gutwein Seeds), (Hoegemeyer Hybrids), (Hoffman Seeds), (Interstate Seed), (Merschman Seeds), (Midwest Seed Genetics), (NC+ Hybrids), (Pioneer

International), (Rupp Seeds), (Sands of Iowa), (Stine Seed) and (Wilson Seeds), 'Liberty Link Field Corn plus *Bt* (Lightning)' (Garst Seed), 'Liberty-Link Independence Canola' (Bayer), 'Liberty-Link Innovator Canola' and 'Topas' (Bayer), 'SeedLink' and 'InVigor' (canola lines MS1 × RF1, MS1 × RF2 and MS8 × RF3)) (Bayer), 'Radicchio Rosso lines RM3-3, RM3-4 and RM3-6' (Bejo Zaden BV), 'Herculex 1' (*cry1F* plus *pat*) (event TC1507) (Dow AgroSciences), (Mycogen) and (Pioneer), 'Herculex 1 plus Roundup Ready Corn 2' (*cry1F* plus *pat* plus *epsps*) (TC1507 plus NK603 (crossed)) (Dow AgroSciences), (Mycogen), (Monsanto) and (Pioneer), 'Herculex Xtra with Roundup Ready Corn 2' (*cry34A(b)1* plus *cry35A(b)1* plus *cry1F* plus *pat* plus *epsps*) (event TC1507 plus DAS 59122-7 plus NK603 (crossed)) (Dow AgroSciences), (Monsanto), (Pioneer) and (DuPont), 'WideStrike' (*cry1A(c)* plus *cry1F* plus *pat*) (Mycogen) and (Dow AgroSciences), 'WideStrike/Roundup Ready' (*cry1A(c)* plus *cry1F* plus *pat* plus *cp4 epsps*) (Mycogen) and (Dow AgroSciences), 'WideStrike/RoundupReady Flex'(*cry1A(c)* plus *cry1F* plus 2 × *pat* plus 2 × *cp4 epsps*) (Dow AgroSciences) and (Pioneer), 'Genuity' and 'SmartStax' (*cry1A.105* plus *cry2A(b2)* plus *cry1F* plus *cry3B(b1)* plus *cry34A(b1)* plus *cry35A(b1)* plus *pat* plus *epsps*) (event MON 88017 plus MON 89034 plus TC1507 plus DAS59122-7) (Monsanto) and (Dow AgroSciences), 'NatureGard' and 'KnockOut' (cry1A(b) plus *pat* – event 176) (Syngenta), 'Agrisure 3100' (*cry1A(b)* plus *pat* – event BT11 × MIR604) (Syngenta), 'Agrisure GT/CB/LL' (*cry1A(b)* plus *cry3A* plus *pat* – event BT11 × GA21) (Syngenta), 'Agrisure CB/LL/RW' (*cry1A(b)* plus *pat* – event BT11 × MIR162 × MIR604 × GA21) (Syngenta), 'Agrisure 3000GT' (*cry1A(b)* plus *cry3A* plus *pat* plus *epsps* – event BT11 × MIR604) (Syngenta), 'NatureGard' and 'KnockOut' (*cry1A(b)* plus *pat* (event 176)) (Syngenta), 'Bt plus Liberty Link Corn' (Garst Seeds).
Patents: EP 275957.

PRODUCT SPECIFICATIONS: The transgenic sugar beet line T120-7 was field tested in the USA (1994, 1996, 1997), Canada, Western and Eastern Europe, and in the former Soviet Union. T120-7 was evaluated extensively and no differences were found in the agronomic characteristics, plant emergence and seedling vigour for line T120-7 compared with non-transformed counterpart beets and standard commercial sugar beet varieties growing in nearby fields. The cotton line 281-24-236 was field tested in the USA from 2000 to 2002 and in Puerto Rico in 2000–01 for a total of 63 location-years. These trials were conducted to evaluate agronomic performance and crop quality, and to determine the plant pest risk potential of 281-24-236. Comparisons were made between the modified line and one of its parental lines, PSC355. Whilst there were small but significant differences in vegetative growth parameters, these were within the observed range of commercial cultivars. Also observed were small but significant differences in some reproductive parameters (such as percentage retention of bolls) and fibre quality. These slight differences would not be expected to confer weediness characteristics to line 281-24-236. The susceptibility of line 281-24-236 to cotton diseases, such as seed rot, *Fusarium* and *Verticillium* wilts, and boll rot, was evaluated and determined similarly to that of the parental line PSC355. Results from these field trials demonstrate that the growth, agronomic performance and disease susceptibility of line 281-24-236 is similar to that of conventional cotton. This cotton line is, therefore, not expected to become a plant pest risk, either in terms of weediness, or in becoming a more suitable host for plant diseases.

COMPATIBILITY: 'Liberty Link' varieties of crops are tolerant of over-the-top applications of glufosinate-ammonium, sold by Bayer CropSciences as 'Liberty'.

MAMMALIAN TOXICITY: There is no evidence that 'Liberty Link' crops have any unusual toxic characteristics that will render them different from conventional crops. There have been no reports of allergic or other adverse effects from researchers, breeders or users of the products. It has been shown that the PAT enzyme has a very low potential for toxic or allergenic effects based on its physicochemical characteristics (such as rapid breakdown under mammalian digestive conditions using simulated gastric and intestinal fluids and lack of heat stability), its low concentrations in plant tissues and thus food or livestock feed products derived from them, and the lack of amino acid sequence homology with any known protein toxins or allergens.

ENVIRONMENTAL IMPACT AND NON-TARGET TOXICITY: There is no evidence that the use of 'Liberty Link' crops will have any deleterious effect on non-target organisms or on the environment. It is unlikely that the gene for glufosinate tolerance will be transferred to other plant species. Volunteers in subsequent crops can be removed by other herbicides. It has been concluded that the *pat* gene inserted into the transgenic crops will not result in any deleterious effects or significant impacts on non-target organisms, including threatened and endangered species or beneficial organisms. Field observations of soybean lines A2704-12, A2704-21 and A5547-35 concluded that they did not have a significant adverse impact on organisms beneficial to plants or agriculture, or on other non-target organisms. These events were not expected to impact on threatened or endangered species as the PAT enzyme is not known to have any toxic properties. The high specificity of the enzyme for its substrates makes it unlikely that the introduced enzyme would metabolise endogenous substrates to produce compounds toxic to beneficial organisms.

DATE(S) OF REGISTRATION: Environmental regulatory approvals for T120-7 were granted in the USA in 1998 and Canada in 2001; food and/or feed approvals were gained in the USA in 1998; food uses approval was received in Japan in 1999 and in Canada in 2000; feed use was granted in Japan in 1999 and Canada in 2001. Environmental regulatory approvals for 'Liberty Link Independence Canola' were achieved in the USA and Canada in 1995 and in Japan in 1997; food and/or feed clearance was granted in the USA in 1995; food approvals were obtained in Canada in 1996 and Japan in 1997; and feed use approval was received in Canada in 1995 and Japan in 1998. Environmental regulatory approvals for 'Liberty Link Innovator Canola' were granted in Canada in 1995, Japan in 1996, the USA in 2002 and Australia in 2003; food and/or feed approvals were granted in Mexico in 1999,

South Africa in 2001 and China in 2004; food use approvals were granted in Canada and USA in 1995, Japan in 1996, Australia in 2001 and Korea in 2005; feed approvals were given in Canada in 1995, Japan in 1996 and Korea in 2005; marketing approvals were granted for the EU in 1998. Environmental regulatory approvals for MS1, RF1 → PGS1 were granted in Canada in 1995, Japan in 1996, the USA in 2002 and Australia in 2003; food and/or feed approval was received in the USA in 1996, South Africa in 2001, Australia in 2002, China in 2004 and the EU in 2005; food use approval was granted in Canada in 1995, Japan in 1996 and Korea in 2005; feed use was cleared in Canada in 1995, Japan in 1996 and Korea in 1998; marketing approval was allowed in the EU in 1996. Environmental regulatory approvals for MS1, RF2 → PGS2 were obtained in Canada in 1995, Japan in 1997, the USA in 2002 and Australia in 2003; food and/or feed uses were approved in the USA in 1996, South Africa in 2001, Australia in 2002, China in 2004 and the EU in 2005; food uses were achieved in Canada in 1995, Japan in 1997 and Korea in 2005; feed uses were approved in Canada in 1995, in Japan in 1997 and Korea in 2005; marketing approval was granted in the EU in 1997. Environmental regulatory approvals for MS8 × RF3 were granted in Canada in 1996, Japan in 1998, the USA in 1999 and in Australia in 2003; food and/or feed approvals were achieved in the USA in 1996, South Africa in 2001, Australia in 2002, China and Mexico in 2004 and the EU in 2005; food uses were approved in Canada and Japan in 1997 and in Korea in 2005; and feed uses were approved in Canada in 1996, Japan in 1998 and Korea in 2005. Japan granted approvals for PHY14 and PHY35 canola for the environment and food uses in 1997 and for feed uses in 1998; and for PHY36 canola environment, food and feed uses were approved in 1997. HCN28 canola was approved for the environment in Canada in 1996, Japan in 1997, the USA in 1998 and Australia in 2003; it received food and/or feed approval in the USA in 1998, Mexico in 2001, Australia in 2002, China in 2004 and the EU in 2009; food approvals were gained in Canada and Japan in 1997 and in Korea in 2005; and feed uses were approved in Canada in 1995, Japan in 1997 and Korea in 2005. HCR-1canola received environmental and feed approval in Canada in 1998. RM3-3, RM3-4 and RM3-6 chicory was approved in the EU in 1996 environmentally and for marketing; environmental approval and use as food and/or feed was achieved in the USA in 1997. Environmental regulatory approvals for A2704-12, A2704-21 and A5547-35 soybeans were achieved in the USA in 1996, Canada and Japan in 1999 and Brazil in 2010; food and/or feed approvals were received in the USA in 1998, South Africa in 2001, Mexico in 2003, China in 2007 and the EU in 2008; food use approvals were achieved in Canada in 2000, Japan in 2002, Australia in 2004, Taiwan in 2007, Korea and the Philippines in 2009 and Brazil in 2010; and feed use approvals were granted in Canada in 2000, Japan in 2003, Korea and the Philippines in 2009 and Brazil in 2010. Environmental regulatory approvals for A5547-127 were granted in the USA in 1998, Canada in 2000, Japan in 2006 and Brazil in 2010; food and/or feed use was approved in the USA in 1998 and Mexico in 2003; food use was allowed in Canada in 2000, Japan in 2003 and Brazil in 2010; and feed approvals were received in Canada in 2000, Japan in 2006 and Brazil in 2010. Soybean line GU262 received environmental and food and/or feed approval in the USA in 1998. 'WideStrike' (*cry1A(c)*, *cry1F* and *pat* genes) was registered as safe for the environment in the USA in 2004 and in Brazil in 2009. It was registered for use in food and/or feed in Mexico and the USA in 2004, in Japan in 2005 and in Brazil in 2009. It was approved for use in food in Australia and Korea in 2005 and for feed in Korea in 2008. Cotton 281-24-236 3006-210-23 (*cry1A(c)*, *cry1F* and *pat* genes) is undergoing evaluation. Regulatory approvals for MON89034, TC1507, MON88017, and DAS-59122-7 (*pat*, *CP4 epsps*, *cry1A.105*, *cry2Ab*, *cry3Bb1*, *cry34Ab1*, *cry35Ab1* and *cry1Fa2*) were granted for environmental impact in Canada, Japan and the USA in 2009; for food and/or feed in Mexico and the Philippines in 2010; and for food and use as feed in Japan, Korea and Taiwan in 2009. Regulatory approvals for 176 (*cry1A(b)* and *pat* genes) were granted for environmental impact in the USA in 1995, Argentina, Canada and Japan in 1996 and the EU in 1997; for food and/or feed in the USA in 1995, in Australia and South Africa in 2001 and in China in 2004; for food in Canada in 1995, in Japan in 1996, in the EU, the Netherlands, Switzerland and the UK in 1997, in Argentina in 1998, in Korea and the Philippines in 2003 and in Taiwan in 2004; and for use as feed in Canada and Japan in 1996, in the EU, the Netherlands and Switzerland in 1997, in Argentina in 1998, in the Philippines in 2003 and in Korea in 2006. It is not approved for cultivation in Japan or Switzerland, and in the EU it was notified as an existing product on 4 October 2004. Notification expired on 18 April 2007 and there has been no application for renewal. Hence the authorisation is no longer valid. Regulatory approval for BT11 (*cry1A(b)* and *pat* genes) in the environment was granted in Canada, the USA and Japan in 1996, in Argentina in 2001, in South Africa in 2003, in Uruguay in 2004, in the Philippines in 2005, in Brazil in 2007 and in Colombia in 2008; food and/or feed use approvals were granted in the USA in 1996, in Australia in 2001, in South Africa in 2002, in China and Colombia in 2004 and in Brazil and Mexico in 2007; food approvals were granted in Canada and Japan in 1996, in the EU, Switzerland and the UK in 1998, in Argentina in 2001, in Korea and the Philippines in 2003, in Taiwan in 2004 and in Colombia in 2008; approvals for use as feed were granted in Canada and Japan in 1996, in the EU, Switzerland and the UK in 1998, in Argentina in 2001, in the Philippines in 2003, in Korea in 2006 and in Colombia in 2008. It is not approved for cultivation in Japan and the EU, but in 1998 field maize and sweetcorn were approved for import and feed use and field maize was approved for food use. In 2004, sweetcorn was approved for food use. Regulatory approval for BT11 × MIR162 (*pat*, *vip3Aa20* and *cry1Ab*) in the environment was allowed in the USA in 2009. Registration of BT11 MIR162 maize was granted by US-EPA under Section 3(c)(7)(C) of FIFRA (EPA Reg. Number: 67979-12). The registration will automatically expire at midnight on 31 December 2011. Registration of BT11 MIR162 MIR604 (*pat*, *cry3A*, *cry1Ab* and *vip3Aa20*) maize in the environment was granted by US-EPA under Section 3(c)(7)(C) of FIFRA (EPA Reg. Number: 67979-13) in 2009. The subject registration will automatically expire at midnight on 31 December 2011.

This product did not require a determination of non-regulated status by USDA-APHIS or a review by US-FDA. Regulatory approvals for BT11 × MIR604 (*pat*, *cry3A* and *cry1Ab*) were given for the environment in Canada in 2007, for food and/or feed in Mexico 2007, for food in Japan, Korea and the Philippines in 2007, and for feed in Korea and the Philippines in 2007. Regulatory approvals for BT11 MIR604 GA21 (*pat*, *epsps*, *cry3A* and *cry1Ab*) for the environment in Canada in 2007 and in Korea in 2008, for food and/or feed in Mexico in 2008 for food in Japan in 2007 and in Korea and the Philippines in 2008, and for use as feed in the Philippines in 2008.

1.2.11 *sethoxydim tolerance* gene

Introduces tolerance to sethoxydim

NOMENCLATURE: Approved name: *sethoxydim tolerance* gene.

Other names: *SR-tolerance* gene; Poast Protected gene.

Development code(s): DK404SR.

Promoter(s): Not relevant – selected trait not transgenic.

Trait introduction: The maize hybrid DK404SR was developed to allow the use of sethoxydim, the active ingredient in the herbicide 'Poast', for the control of annual and perennial grasses in maize crops. DK404SR originated from a somaclonal variant selected from embryo tissue and expresses a modified version of the acetyl-CoA-carboxylase (ACCase) enzyme.

SOURCE: A collaborative programme between BASF and DeKalb Genetics was established to identify maize plants that showed resistance to post-emergence applications of sethoxydim. The collaboration identified several plants that were used in conventional breeding programmes to produce 'SR' hybrids that combined tolerance to sethoxydim with the quality traits of elite maize varieties.

PRODUCTION: The maize hybrid DK404SR was developed from sethoxydim-tolerant inbred lines. The original sethoxydim-tolerant mutant line (S2) was selected as a somaclonal variant from maize embryo tissue grown under sethoxydim-selective pressure. The process involved growing somatic embryos on sethoxydim-enriched media. From the somatic embryos that survived, the somaclonal variant cell line S2 was selected and subsequently regenerated. This S2 line was backcrossed at least six times with both parental lines of the hybrid DK404SR to transfer the sethoxydim-tolerant trait.

TARGET PESTS: Grass weeds.

TARGET CROPS: Maize (corn).

BIOLOGICAL ACTIVITY: The cyclohexanedione herbicides show herbicidal effects against grass weeds, and this includes graminaceous crops such as maize. The selected maize tolerant of sethoxydim can be treated with over-the-top applications of the herbicide without damage. **Mode of action:** ACCase is a key enzyme in the fatty acid biosynthesis pathway, necessary for the synthesis and maintenance of membranes and for the incorporation of fatty acids into triacylglycerides. The enzyme is inhibited by cyclohexanedione herbicides, such as sethoxydim, resulting in a lethal disruption of lipid biosynthesis. The mutant *ACCase* gene encodes for a modified version of acetyl-CoA-carboxylase enzyme in which the sethoxydim-binding site is altered such that the ACCase enzyme is not inhibited by the sethoxydim herbicide, yet still retains its normal catalytic properties in fatty acid synthesis. The modified maize plants also demonstrated some level of cross-tolerance to the related herbicides, aryloxyphenoxypropionates. The basis of the tolerance is not certain, but it may be associated with a change in the structure of the ACCase enzyme, the target for this herbicide.

COMMERCIALISATION: Formulation: BASF has been working with DeKalb Genetics with a view to launching sethoxydim-tolerant maize into the US market. A number of seed companies are selling hybrids in the USA. **Trade names:** 'Poast Protected Corn' (withdrawn) (Asgrow Seed), (Cargill Hybrid Seeds), (Countrymark Cooperative), (Croplan Genetics), (DeKalb) and (Growmark).

COMPATIBILITY: The sethoxydim-tolerant maize hybrids can be sprayed with full field rates of sethoxydim as 'Poast' and are compatible with normal treatments used on maize.

MAMMALIAN TOXICITY: Sethoxydim-tolerant maize is considered to be substantially identical to conventional varieties and is not considered to pose any mammalian risks. There have been no reports of allergic or other adverse toxicological effects from research workers, breeders or field staff.

ENVIRONMENTAL IMPACT AND NON-TARGET TOXICITY: The breeding lines for these hybrids were selected from conventionally bred maize and are not considered to show any adverse effects on non-target organisms or on the environment.

DATE(S) OF REGISTRATION: 'Poast Protected Corn' was approved for the environment and feed use in 1996 and for food use in 1997 in Canada. It is no longer registered.

1.3 Insect resistance

1.3.1 *Bacillus thuringiensis crylA* endotoxin genes

Introduces resistance to insects

NOMENCLATURE: Approved name: *Bacillus thuringiensis* Berliner *crylA* endotoxin gene; *cry1A.105* gene.

Other names: *Bt* gene; *crylA(b)*; *crylA(c)*; chimeric *cry1* delta-endotoxin; *cry1A.105*.

Development code(s): SYN-BTØ11-1 (BT11 (X4334CBR, X4734CBR)); BT11 × MIR162 × MIR604; SYN-BTØ11-1, SYN-IR6Ø4-5 (BT11 × MIR604); SYN-BTØ11-1, SYN-IR6Ø4-5, MON-ØØØ21-9 (BT11 × MIR604 × GA21); SYN-BTØ11-1, SYN-IR162-4 (BT11 × MIR162); MON802; MON809; MON-ØØ81Ø-6 (MON810); MON-ØØ863-5, MON-ØØ81Ø-6 (MON863 × MON810); MON-ØØ6Ø3-6, MON-ØØ81Ø-6 (NK603 × MON810); MON-ØØØ21-9, MON-ØØ81Ø-6 (GA21 × MON810); REN-ØØØ38-3, MON-ØØ81Ø-6 (MON810 × LY038); ACS-ZMØØ3-2,MON-ØØ81Ø-6 (T25 × MON810); MON-ØØ863-5, MON-ØØ81Ø-6, MON-ØØ6Ø3-6 (MON863 × MON810 × NK603); MON-88Ø17-3, MON-ØØ81Ø-6 (MON810 × MON88017); MON80100; MON88017 × DAS-59122-7); MON-89Ø34-3 (MON89034); MON-89Ø34-3, MON-88Ø17-3 (MON89034 × MON88017); MON-89Ø34-3, MON-ØØ6Ø3-6 (MON89034 × NK603); MON-89Ø34-3 × DAS-Ø15Ø7-1 × MON-88Ø17-3 × DAS-59122-7 (MON89034 × TC1507 × SYN-EV176-9 (176); DAS-21Ø23-5 × DAS-24236-5; DAS-21Ø23-5 × DAS-24236-5; 281-24-236 × 3006-210-23; MON 531; MON531 × MON1445; MON15985; LL25 × MON15985; MON15985 × MON1445; MON88913 × MON15985; MON87701; MON87701 × MON89788.

Promoter(s): *Cry1A.105* – CaMV 35S 5′ untranslated leader from wheat chlorophyll *a-/b*-binding protein rice actin gene intron. *Cry1A(b)* – first copy promoter from the maize phosphoenolpyruvate carboxylase gene, the CaMV 35S terminator (the second copy under the regulation of a promoter derived from a maize calcium-dependent protein kinase gene) and the CaMV 35S terminator. The Bt-maize Event 176 produces a truncated version of the insecticidal protein, Cry1Ab, derived from *Bacillus thuringiensis* subsp. *kurstaki* strain HD-1. Event 176 is also genetically modified to express the *bar* gene cloned from the soil bacterium *Streptomyces hygroscopicus* (Jensen) Waksman & Henrici, which encodes a phosphinothricin-*N*-acetyltransferase (PAT) enzyme and is driven by CaMV 35S′. The PAT enzyme is useful as a selectable marker enabling identification of transformed plant cells, as well as a source of resistance to the herbicide phosphinothricin (also known as glufosinate ammonium, the active ingredient in the herbicides 'Basta', 'Rely', 'Finale' 'Harvest' and 'Liberty' (see *The Pesticide Manual* Fifteenth Edition, entry 444)). PAT catalyses the acetylation of phosphinothricin, and thus detoxifies phosphinothricin, eliminating its herbicidal activity. Constitutive expression of the *cry1Ab* gene in BT11 is controlled by the 35S promoter derived from cauliflower mosaic virus (CaMV). The *cry1A(c)* gene in DAS-24236-5 × DAS-21Ø23-5 is controlled by the maize ubiquitin 1 promoter.

Trait introduction: Trait introduction into MON80100, 176 and MON15985 was by microparticle bombardment of plant cells or tissue. Event 176 was produced by biolistic transformation of the inbred line CG00526 with two plasmids. One plasmid contained two copies of a 3′ truncated *cry1Ab* gene, each regulated by different promoter sequences. The *cry1Ab* open reading frame, corresponding to the sequence encoding the *N*-terminal 648 amino acids of the native Cry1Ab protein, was modified for optimal expression in plant cells. Green tissue expression of one copy of the *cry1Ab* gene was regulated by the phosphoenolpyruvate carboxylase promoter while expression of the other *cry1Ab* gene was controlled by a pollen-specific promoter isolated from maize. Both genes employed 3′-polyadenylation sequences from the 35S transcript of cauliflower mosaic virus (CaMV). *Cry1A.105*, MON531 and MON87701 trait introduction was by *Agrobacterium tumefaciens* (Conn.)-mediated plant transformation. Trait introduction to BT11 × MIR162, BT11 × MIR162 × MIR604, BT11 × MIR604 × GA21, BT11 × MIR604, DAS-24236-5 × DAS-21Ø23-5, MON89034 × NK603 and MON89034 × TC1507 × MON88017 × DAS-59122-7 was by traditional plant breeding and selection. The BT11 corn line was created through direct DNA transformation of plant protoplasts from the inbred maize line H8540 and regeneration on selective medium. A single plasmid, designated pZO1502, was used in the transformation event and contained a truncated synthetic *cry1Ab* gene encoding Cry1Ab endotoxin and a synthetic *pat* gene (to allow transformant selection on glufosinate ammonium). The *cry2A(b)* gene was introduced into the cotton line DP50B already containing the *cry1A(c)* gene. As a result, event 15985 expresses both the Cry1Ac and Cry2Ab insecticidal proteins.

SOURCE: *Bacillus thuringiensis* Berliner subsp. *kurstaki* (*Btk*) produces parasporal, proteinaceous, crystal inclusion bodies during sporulation. Upon ingestion, these are insecticidal to larvae of the order Lepidoptera and to both larvae and adults of a few Coleoptera. Monsanto has used *cry1A(c)* genes in cotton and tomatoes and *crylA(b)* and *cry1A.105* genes (all from *Btk*) in maize (corn); Novartis (now Syngenta) and Mycogen (now Dow AgroSciences) have used *cry1A(b)* genes from *Btk*, also in maize. In the USA, the EPA must register events before they can be used in commercial seed production. The EPA has registered several unique *Bt cry1A* events including: 176 (Novartis Seeds (now Syngenta Seeds) and Mycogen), BT11 (Northrup King/Novartis Seeds), MON 810 (Monsanto) and DBT418 (DeKalb Genetics). The MON 810 and BT11 events are used in production of 'YieldGard' corn. Event 176 is sold as 'KnockOut' by Syngenta and 'NatureGard' by Dow AgroSciences. The DBT418 event is sold as 'BT-Xtra'.

PRODUCTION: Isolated from *Bacillus thuringiensis* Berliner subsp. *kurstaki*. The *cry1A.105* gene is a chimeric gene comprising four domains from other *cry* genes previously used in transgenic plants. The amino acid sequences of Domains I and II are identical with the respective domains from Cry1Ab and Cry1Ac proteins, Domain III is almost identical to the Cry1F protein, and the C-terminal Domain is identical to Cry1Ac protein. The gene is isolated from *Bacillus thuringiensis*, is often truncated and introduced into the crop associated with the promoter, usually cauliflower mosaic virus 35S promoter. Before introduction into crops it is common to increase the number of constructs in bacterial fermentation. The transgenic bacteria may be identified by use of a selectable marker such as antibiotic-resistance linked to a bacterium-specific promoter. The plants are produced by insertion of the *Bt* nucleic acid using transformed and disabled *Agrobacterium tumefaciens*, or bombardment using a particle gun or other accepted transformation techniques. Transgenic crops are identified using a selectable marker such as a herbicide-tolerance or an antibiotic-resistance gene. Elite varieties are bred from these transgenic crops. Syngenta (as Novartis) launched *Bt* 176, a transgenic insect-resistant maize that contained two selectable markers, glufosinate-tolerance and ampicillin-resistance. More recently, it has developed a second maize line, *Bt* 11, that contains no antibiotic-resistance genes. **Gene flow:** The crops transformed with *Bt* genes will not cross-pollinate indigenous weed species and gene flow, and the possible build-up of insect resistance through contact with non-crop species is very unlikely.

TARGET PESTS: Lepidoptera including European cornborer (ECB) (*Ostrinia nubilalis* Hübner), pink cornborer (*Sesamia cretica* Lederer) and noctuids (*Heliothis* spp. and *Helicoverpa* spp.). The *cry1A*(*c*)-gene encodes the Bt-toxin Cry1A(c), which confers resistance to lepidopteran pests of cotton, such as tobacco budworm (*Heliothis virescens* (Fabricius)), cotton bollworm (*Helicoverpa zea* Boddie), pink bollworm (*Pectinophora gossypiella* (Saunders)) and soybean looper (*Pseudoplusia includens* (Walker)).

TARGET CROPS: The main target crops are cotton and maize (corn). There are some soybean crops containing *cry1A* genes.

BIOLOGICAL ACTIVITY: The gene is expressed constitutively throughout the plant such that susceptible insects that feed upon the crop succumb. **Biology:** The *cry1A*(*b*) gene codes for the full endotoxin that is hydrolysed by the gut digestive enzymes to form the toxin. It is active against a wide variety of Lepidoptera and is used to transform maize to give resistance to European cornborer (*O. nubilalis*). The *cry1A*(*c*) gene codes for the endotoxin that is particularly active against tobacco budworm (*H. virescens*), a major pest of cotton. Many crops are now being introduced that contain both *Bt* genes and herbicide-tolerance genes.
Mode of action: Once in the insect, the crystal proteins are solubilised and the insect gut proteases convert the original pro-toxin into a smaller toxin. These hydrolysed toxins bind

to the insect's midgut at high-affinity, specific receptor-binding sites, where they interfere with the potassium ion-dependent active amino acid symport mechanism. This disruption causes the formation of a large cation-selective pore that increases the water permeability of the membrane. A large uptake of water causes epithelial cell swelling and eventual rupture, with the subsequent disintegration of the midgut lining. Different toxins bind to different receptors, and this explains the selectivity of different *Bt* strains. (This differential binding is described by some of those working in the field as different modes of action.) Different genes coding for these protein toxins have been isolated and used to transform various crop plants such that the crops produce insecticidal proteins throughout their life. In some cases, the truncated genes have been used so that the insect-active toxin is produced in the crop rather than the pro-toxin that requires hydrolysis in the insect gut. In 'KnockOut' and 'NatureGard', the *Bt*-toxin gene is *crylAb* and tissues that produce the toxin are green pollen and the stalk. The toxin level is relatively high early in the season in green tissue, but in the late season, the toxin levels decline rapidly. 'KnockOut' was developed by Ciba Seeds (then Novartis and now Syngenta) and Mycogen (now Dow AgroSciences). The genetic event name is 176. In 'Bt-Xtra', the *Bt*-toxin gene is also *crylA*(*b*), but the tissues that produce the toxin are leaf, kernel, stalk and silk. The toxin level is high in the leaves full season, but with much lower levels in kernel, stalk and silk. 'Bt-Xtra' was developed by DeKalb. The genetic event names are DeKalBt and DBT418. In 'YieldGard', the *Bt*-toxin gene is *crylA*(*c*). The tissues that produce the pollen are leaf, pollen, tassel, silk and kernel. The toxin level is high in leaves, pollen, tassel, silk and kernel for the whole season. 'YieldGard' was developed by Monsanto and Northrup King (Novartis – now Syngenta) and is marketed by Pioneer, Cargill, DeKalb and Golden Harvest. The genetic event names are MON 810 and Bt11. **Efficacy:** Field studies have shown good control of the target insects in the absence of additional chemical applications. Where other insect pests, especially non-lepidopteran pests, have appeared in the crop, additional chemical treatment is usually recommended and often necessary. **Key reference(s):** 1) Y Bertheau, JC Helbling, MN Fortabat, S Makhzami, I Sotinel, C Audeon, AC Nignol, A Kobilinsky, L Petit, P Fach, P Brunschwig, K Duhem & P Martin. 2009. Persistence of plant DNA sequences in the blood of dairy cows fed with genetically modified (Bt176) and conventional corn silage. *J. Agric. Food Chem.*, **57**, 509-16. 2) FS Betz, BG Hammond & RL Fuchs. 2000. Safety and advantages of *Bacillus thuringiensis*-protected plants to control insect pests. *Regulatory Toxicology*, *32*, 156-73. 3) Y Li, M Meissle & J Romeis. 2008. Consumption of Bt maize pollen expressing Cry1Ab or Cry3Bb1 does not harm adult green lacewings, *Chrysoperla carnea* (Neuroptera: Chrysopidae). *PLoS ONE*, **3(8)**, e2909. 4) AP Oliveira, ME Pampulha & JP Bennett. 2008. A two-year field study with transgenic *Bacillus thuringiensis* maize: effects on soil microorganisms. *Sci. Total Environ.*, **405**, 351-7. 5) J Brake & D Vlachos. 1998. Evaluation of transgenic event 176 'Bt' corn in broiler chickens. *Poultry Science*, **77**, 648-53. 6) X Wu, LB Rogers, YC Zhu, CA Abel, GP Head & F Huang. 2009. Susceptibility of Cry1Ab-resistant and -susceptible sugarcane borer

(Lepidoptera: Crambidae) to four *Bacillus thuringiensis* toxins. *J. Invertebr. Pathol.*, **100(1)**, 29-34. 7) JJ Adamczyk, SM Greenberg, JS Armstrong, WJ Mullins, LB Braxton, RB Lassiter & MW Siebert. 2008. Evaluations of Bollgard, Bollgard II, and WideStrike technologies against beet and fall armyworm larvae (Lepidoptera: Noctuidae). *Florida Entomologist*, **91(4)**, 531–6. 8) MD Dryzga, BL Yano, AK Andrus & JL Mattsson. 2006. Evaluation of the safety and nutritional equivalence of a genetically modified cottonseed meal in a 90-day dietary toxicity study in rats. *Food Chem. Toxicol.*, **45**, 1994–2004.

COMMERCIALISATION: Formulation: Approval for Monsanto's transgenic cotton seed (line 531) was granted in the USA in late 1995, with the first introduction of 'BollGard' cotton in 1996 by Delta and Pineland as 'NUCOTN 33B' and 'NUCOTN 35B'. Monsanto's DBT 418 maize, containing the *Bacillus thuringiensis* var *kurstaki cry1A(c)* gene for resistance to European cornborers and the phosphinothricin acetyl transferase (*pat*) gene for tolerance to the herbicide glufosinate-ammonium, was introduced in 2001. Ciba (now Syngenta) was granted approval to sell *Bt*-transformed maize in the USA in 1995 and in Canada in 1996. Monsanto has a tomato transformed with *cry1Ac* and registered in the USA in March 1998, but it is not commercialised. **Trade names:** 'NUCOTN 33B' and 'NUCOTN 35B' (*cry1A(c)* – event 531) (Monsanto), 'BollGard Cotton' (*cryIA(c)* – events 531, 757, 1076) (Monsanto), (Deltapine), (Paymaster Cottonseed) and (Stoneville Pedigreed Seed), 'BollGard II' (*cry1A(c)* and *cryIIA(b2)* – event 15985) (Monsanto), 'InGard Cotton' (*cryIA(c)* – events 531, 757, 1076) (Deltapine Australia) and (Monsanto), 'BollGard plus Roundup Ready Cotton' (*cryIA(c)*) (Deltapine) and (Paymaster Cottonseed), 'Bt plus Buctril BXN System Cotton' (*cryIA(c)* plus *bxn* – events 31807 and 31808) (Monsanto) and (Stoneville Pedigreed Seed), 'WideStrike Cotton' (*cry1A(c)* plus *pat* – event 15985 (DAS-21Ø23-5 × DAS-24236-5)) (Dow AgroSciences (Mycogen)) 'YieldGard Corn' (*cry1A(b)* – event *Bt* MON 810) (Monsanto), (Asgrow Seeds), (Becks Hybrids), (Cargill Hybrid Seeds), (Countrymark Cooperative), (Croplan Genetics), (DeKalb Genetics), (Golden Harvest Seeds), (Growmark), (Hoffman Seeds) and (Pioneer), 'YieldGard Corn Borer Corn with Roundup Ready 2' (*cry1A(b)* plus *epsps* – event *Bt* MON 810 × NK603) (Monsanto), 'YieldGard Plus with Roundup Ready Corn 2' (*cry1A(b)* plus *cry3B(b1)* plus *epsps* – event *Bt* MON 810 × MON 863 × NK603) (Monsanto), 'YieldGard VT Triple Rootworm Roundup Ready 2 Corn' (*cry1A(b)* plus *cry3B(b1)* plus *epsps* – event *Bt* MON 810 × MON 88017) (Monsanto), 'NatureGard' and 'KnockOut' (*cry1A(b)* plus *pat* – event 176) (Syngenta), 'Maximizer Corn' (*cryIA(b)*), 'KnockOut', 'MaizeGard Corn' and 'MaisGard Corn' (Syngenta), 'Bt plus Liberty Link Corn' (Garst Seeds), 'Bt Corn plus IMI Tolerance' (Garst Seeds), (Dow AgroSciences (Mycogen)) and (Syngenta), 'Bt-Xtra Corn' (DeKalb Genetics), 'Agrisure 3100' (*cry1A(b)* plus *pat* – event BT11 × MIR604) (Syngenta), 'Agrisure CB/LL' (*cry1A(b)* plus *pat* – event BT11) (Syngenta) , 'Agrisure GT/CB/LL' (*cry1A(b)* plus *cry3A* plus *pat* – event BT11 × GA21) (Syngenta), 'Agrisure CB/LL/RW' (*cry1A(b)* plus *pat* – event BT11 × MIR162 × MIR604 × GA21) (Syngenta) , 'Agrisure 3000GT' (*cry1A(b)* plus *cry3A* plus *pat* plus *epsps* – event BT11 × MIR604) (Syngenta), 'Genuity VT' and 'Double PRO' (*cry1A.105* plus *cry2A(b2)* plus *epsps* – event MON 89034 × NK603) (Monsanto), 'Genuity VT Triple PRO' (*cry1A.105* plus *cry2A(b2)* plus *cry3B(b1)* plus *epsps* – event MON 89034 × MON 88017), 'Genuity' and 'SmartStax' (*cry1A.105* plus *cry2A(b2)* plus *cry1F* plus *cry3B(b1)* plus *cry34A(b1)* plus *cry35A(b1)* plus *pat* plus *epsps*) (Monsanto) and (Dow AgroSciences). **Patents:** US Patent Application 2008/0260932 A1 (*cry1A.105*).

APPLICATION: Seed expressing the endotoxins from *Bacillus thuringiensis* shows tolerance to phytophagous insects, particularly from the order Lepidoptera. Its use often results in reduced applications of conventional insecticides.

COMPATIBILITY: Compatible with most crop protection agents. Cotton growers are recommended to plant 75% of their crop to *Bt*-cotton and to apply conventional chemicals to the remaining 25%, which serves as a refuge and should not be treated with *Bacillus thuringiensis*-based products. Alternatively, they may sow 96% of their crop to *Bt*-cotton and leave the remaining 4% completely untreated. Refugia are designed to reduce the possibility of the onset of insect resistance to *Bt* endotoxins. In maize, a refuge of 20% of the planted area is required in non-cotton growing regions. This refuge should be located within 0.8 km (half a mile) of the *Bt*-maize. The refuge may be treated if insect damage reaches the economic threshold, but *Bt*-based insecticides must not be used.

MAMMALIAN TOXICITY: The truncated *Bt* CryIA toxin extracted from maize leaf tissue displayed activity that was similar to that produced in *E. coli* modified to produce CryIA endotoxins. The acute oral toxicity test with bacterially derived CryIA protein caused no test substance-related deaths at a dose of 4000 mg/kg. In addition, the endotoxin caused no allergic responses in any animal. There is minimal chance of dermal exposure or exposure by inhalation, as the endotoxins are enclosed within plant cells.

ENVIRONMENTAL IMPACT AND NON-TARGET TOXICITY: A review of the studies submitted for the registration of the *B. thuringiensis* Cry1A produced in maize demonstrated a lack of adverse effects on birds, aquatic invertebrates, honeybee larvae, coccinellid predators and earthworms. Studies were conducted on several non-target organisms to determine the potential toxic effects of Cry1Ab protein on test organisms including adult honeybees (*Apis melifera* L.), a predator ladybird (*Coleomegilla maculate* (DeGeer)), juveniles of the soil-dwelling invertebrate Collembola (springtails, *Folsomia candida* Willem), earthworms (*Eisenia foetida* (Savigny)), juveniles of the freshwater invertebrate *Daphnia magna* Straus, fall armyworm (*Spodoptera frugiperda* JE Smith) and black cutworm, (*Agrotis ipsilon* Hufnagel), northern bobwhite chicks (*Colinus virginianus* L.), and mice (*Mus musculus* L.). Additional studies assessed the impact of Cry1Ab protein on the relative abundance of beneficial arthropods. Results demonstrated that the larval development of honeybees and ladybirds were not affected when reared on pollen collected from Event 176 maize plants as compared with pollen from non-transgenic plants. Similarly, there was no effect on survival, immobilisation, or sublethal toxicity reported for the small aquatic insect, *Daphnia magna* when exposed to pollen collected from Event 176. Survival rates, signs of toxicity, or loss of weight were not observed in earthworms exposed to leaf tissue from Event 176 maize as compared with the control treatments. Two lepidopteran insects (fall armyworm and black cutworm) that are not susceptible to native Cry1Ab were likewise not affected when feed Cry1Ab was derived from leaf tissues. Three insects (ECB, corn earworm (*Helicoverpa zea* (Boddie)) and cabbage looper (*Trichoplusia ni* Hübner)) that are susceptible to native Cry1Ab were also susceptible to plant-produced Cry1Ab. Results from high-dose feeding studies of bobwhite quail fed a protein extract enriched in Cry1Ab isolated from Event 176 maize demonstrated no adverse effects on the birds. Negative effects were observed in Collembola fed with Event 176 leaf protein (5% mortality at 0.088 mg Cry1Ab per kg soil) while collembolans fed with non-Bt maize protein were not adversely affected. The level of Cry1Ab was approximately 10 times greater than the maximum soil concentration that would occur if plants were to be incorporated into the soil at anthesis. Under normal conditions, maize would be ploughed into the soil in the autumn when plants have senesced, and the Cry1Ab protein would be present at very low levels at this time. **Bird toxicity:** When administered by oral gavage at a dose of 2000 mg protein/kg body weight, *Bt*-maize had no apparent effect upon bobwhite quail after 14 days. Thus, the acute toxicity LD_{50} value for quail was determined to be >2000 mg protein/kg body weight. In view of the lack of acute toxicity with *Bt*-maize, no avian hazard is expected from the proposed uses of this plant-pesticide. **Fish toxicity:** The requirement for a static renewal toxicity study has been waived, based on a lack of exposure of fish to the *Bt* endotoxin Cry1A protein produced in maize.

Other aquatic toxicity: In a 48-h static renewal toxicity study of maize pollen containing *Bt* Cry1A delta-endotoxin, *Daphnia magna* was treated at five concentration levels, including a maximum hazard dosage of 150 mg/litre (nominal) of water. No mortalities were observed at any of the treatment levels tested. Thus, the 48-h EC_{50} was determined to be >150 mg/litre. The LOEC (lowest observed effect concentration) and NOEC (no observed effect concentration) were found to be 150 mg/litre.

Effects on beneficial insects: Maize pollen containing the Cry1A toxin caused no significant adverse effects in either ladybirds or earthworms. **Monarch butterflies:** Reports of toxicity of high doses of *Bt* toxin in pollen falling onto milkweed plants to monarch butterfly (*Danaus plexippus* (L.)) larvae in the laboratory do not translate into exposure to toxic levels in the field. Further, the monarch butterfly is neither an endangered nor a threatened species. It is an abundant and widespread insect which, in North America, ranges from central Mexico to southern Canada. There are many factors that cause severe mortality of monarch butterflies, among these are predation, parasitism, destruction of the overwintering habitat and, most notably, climatic variations. Nevertheless, the US-EPA has issued the following warning to farmers – 'The potential for non-target species (e.g.

monarch butterfly larvae) to be affected by *Bt* corn pollen remains under study. As an interim measure, the EPA is encouraging growers to place the non-Bt corn refuge between *Bt* corn and habitats such as prairies, forests, conservation areas and roadsides.' **Persistence in the environment:** *Bacillus thuringiensis*-derived Cry IA protein, when added to soil as a component of maize tissue, decreased, with an estimated half-life (DT_{50}) of 15 days. When incubated without soil, it had a DT_{50} of 25.6 days and a DT_{90} of 40.7 days. The bioactivity of purified Cry IA protein in soil decreased, with an estimated DT_{50} of 8.3 days and a DT_{90} of 32.5 days. Small-scale field studies demonstrated that the number of insects and insect diversity observed on plots planted to either *cry1(c)*-derived maize or a non-transformed counterpart were not significantly different. However, when compared with insect populations on plants treated with a common chemical insecticide (permethrin) versus *cry1(c)* maize plants, the total numbers of beneficial insects (especially ladybirds) associated with GM-maize plants were higher. **Weediness potential:** No competitive advantage was conferred to GM-maize, other than that conferred by resistance to phytophagous insects. Resistance to these insect pests will not, in itself, render maize weedy or invasive of natural habitats since none of the reproductive or growth characteristics were modified. Cultivated maize is unlikely to establish in non-cropped habitats and there have been no reports of maize surviving as a weed. In agriculture, maize volunteers are not uncommon, but are easily controlled by mechanical means or by using herbicides. *Zea mays* L. is not invasive and is a weak competitor with very limited seed dispersal.

DATE(S) OF REGISTRATION: Cotton: 'WideStrike' (*cry1A(c)*, *cry1F* and *pat* genes) was registered as safe for the environment in the USA in 2004 and in Brazil in 2009. It was registered for use in food and/or feed in Mexico and the USA in 2004, in Japan in 2005 and in Brazil in 2009. It was approved for use in food in Australia and Korea in 2005 and for feed in Korea in 2008. Cotton 281-24-236 × 3006-210-23 (*cry1A(c)*, *cry1F* and *pat* genes) is undergoing evaluation. MON15985 (*cry1A(c)* and *cry2A(b)* genes) was placed on the market in the USA on 1 January 2003, but applications have been withdrawn. The European Food Safety Authority (EFSA) received the applications on 6 January 2005 and they were accepted on 16 September 2005. However, additional information has been requested and the proceedings have been halted. An application for renewal of an existing product according to Regulation 1829/2003 on genetically modified food and feed was requested for MON 15985 101 × MON 1445 (*crylA(c)*, *cry2A(b2)* plus *cp4 epsps* genes). This was checked for completeness and accepted on 9 April 2008 by EFSA. However, additional information was requested from the applicant and the proceedings have been suspended. An application for renewal of an existing product according to Regulation 1829/2003 on genetically modified food and feed was made for MON 531 (*cry1A(c)*). It was checked for completeness and accepted on 11 June 2008 by EFSA. However, additional information has been requested from the applicant and the proceedings have been suspended. An application for renewal of an existing product according to Regulation 1829/2003 on genetically modified food and feed was made for MON531 × MON1445 (*crylA(c)* and *cp4 epsps* genes) on 17 April 2007. This was checked for completeness and accepted on 12 March 2008, but additional information has been requested from the applicant and the proceedings have been suspended. An application for MON15985 011 × MON88913 (*cryla(c)*, *cry2a(b2)* and *cp4 epsps* genes) was made in April 2007 and checked for completeness by the EFSA. Additional information has been requested from the applicant and the proceedings have been suspended. An application for MON87701 (*cry1A(c)*) was submitted on 17 May 2010, checked for completeness and accepted on 11 June 2010 by EFSA. Additional information has been requested from the applicant and the proceedings have been suspended. An application was submitted for the hybrid MON87701 × MON89788 (*cry1A(c)* and *cp4 epsps*-genes) for import and processing and for use as food and feed on 27 August 2007. It was checked for completeness and accepted on 8 December 2009 by EFSA. However, additional information has been requested from the applicant and the proceedings have been suspended. **Maize:** Environmental impact applications were approved for MON89034 maize (*cry1A.105* and *cry2Ab*) in the USA, Japan and Canada in 2008 and in Brazil in 2009. It was approved for food and/or feed in the USA in 2007 and in the EU and Brazil in 2009. It was cleared for food use in Japan in 2007, in Australia, Canada, and Taiwan in 2008 and in the Philippines in 2009. It was approved for use as feed in Canada, Colombia and Japan in 2008 and in Korea and the Philippines in 2009. Food and/or feed approvals for MON89034 × MON88017 (*cry1A.105*, *cry2Ab(2)*, *cry3Bb(1)* and *CP4 epsps*-genes) were granted in Japan in 2008, in the Philippines in 2009 and in Mexico in 2010. It was approved for food use in Korea and Taiwan in 2009 and for use as feed in Taiwan in 2009. Regulatory approvals for MON89034 × NK603 (*cry1A.105*, *cry2Ab*, *CP4 epsps* (rice actin 1 promoter) and *CP4 epsps* (enhanced CaMV 35S)-genes) were granted for food and/or feed in Japan in 2008 and in Mexico in 2010, for food in the Philippines and Taiwan in 2009 and in Korea in 2010, and for use as feed in Korea and the Philippines in 2009. Regulatory approvals for MON89034 × TC1507 × MON88017 × DAS-59122-7 (*pat*, *CP4 epsps*, *cry1A.105*, *cry2Ab*, *cry3Bb1*, *cry34Ab1*, *cry35Ab1* and *cry1Fa2*) were granted for environmental impact in Canada, Japan and the USA in 2009, for food and/or feed in Mexico and the Philippines in 2010, and for food and use as feed in Japan, Korea and Taiwan in 2009. Regulatory approvals for 176 (*cry1A(b)* and *bar* genes) were granted for environmental impact in the USA in 1995, Argentina, Canada and Japan in 1996 and the EU in 1997; for food and/or feed in the USA in 1995, in Australia and South Africa in 2001 and in China in 2004; for food in Canada in 1995, in Japan in 1996, in the EU, the Netherlands, Switzerland and the UK in 1997, in Argentina in 1998, in Korea and the Philippines in 2003 and in Taiwan in 2004; and for use as feed in Canada and Japan in 1996, in the EU, the Netherlands and Switzerland in 1997, in Argentina in 1998, in the Philippines in 2003 and in Korea in 2006. It is not approved for cultivation in Japan or Switzerland and in the EU it was notified as an existing product on 4 October 2004. Notification expired on 18 April 2007 and there has been no application for renewal. Hence the authorisation is no longer valid. Regulatory approval for BT11 (*cry1A*(*b*) and *pat* genes) in the environment was granted in Canada, the USA and Japan in 1996, in Argentina in 2001, in South Africa in 2003, in Uruguay in 2004, in the Philippines in 2005, in Brazil 2007 and in Colombia in 2008; food and/or feed use approvals were granted in the USA in 1996, in Australia in 2001, in South Africa in 2002, in China and Colombia in 2004 and in Brazil and Mexico in 2007; food approvals were granted in Canada and Japan in 1996, in the EU, Switzerland and the UK in 1998, in Argentina in 2001, in Korea and the Philippines in 2003, in Taiwan in 2004 and in Colombia in 2008; approvals for use as feed were granted in Canada and Japan in 1996, in the EU, Switzerland and the UK in 1998, in Argentina in 2001, in the Philippines in 2003, in Korea in 2006 and in Colombia in 2008. It is not approved for cultivation in Japan and the EU, but in 1998, field maize and sweetcorn were approved for import and feed use and field maize was approved for food use. In 2004, sweetcorn was approved for food use. Regulatory approval for BT11 × MIR162 (*pat*, *vip3Aa20* and *cry1Ab*) in the environment was allowed in the USA in 2009. Registration of BT11 × MIR162 maize was granted by US-EPA under Section 3(c)(7)(C) of FIFRA (EPA Reg. Number: 67979-12). The registration will automatically expire at midnight on 31 December 2011. Registration of BT11 × MIR162 × MIR604 (*pat*, *cry3A*, *cry1Ab* and *vip3Aa20*) maize in the environment was granted by US-EPA under Section 3(c)(7)(C) of FIFRA (EPA Reg. Number: 67979-13) in 2009. The subject registration will automatically expire at midnight on 31 December 2011. This product did not require a determination of non-regulated status by USDA-APHIS or a review by US-FDA. Regulatory approvals for BT11 × MIR604 (*pat*, *cry3A* and *cry1Ab*) were given for the environment in Canada in 2007, for food and/or feed in Mexico in 2007, for food in Japan, Korea and the Philippines in 2007 and for feed in Korea and the Philippines in 2007. Regulatory approvals for BT11 × MIR604 × GA21 (*pat*, *epsps*, *cry3A* and *cry1Ab*) for the environment in Canada in 2007 and in Korea in 2008, for food and/or feed in Mexico in 2008, for food in Japan in 2007 and in Korea and the Philippines in 2008 and for use as feed in the Philippines in 2008. Regulatory approvals for GA21 × MON810 (*cry1Ab* and *CP4 epsps*) were granted for food and/or feed in Japan and South Africa in 2003 and in the EU in 2005; for food in Korea and the Philippines in 2004 and for feed in the Philippines in 2004. It was notified in the EU as an existing product on 6 October 2004. Notification expired on 18 April 2007 with no application for renewal. Therefore, the authorisation is no longer valid. Regulatory approvals for MON80100 (*goxv247*, *cry1Ab* and *CP4 epsps*) were granted in the USA for the environment in 1995 and as food and/or feed in 1996. Regulatory approvals for MON802 (*goxv247*, *cry1Ab* and *CP4 epsps*) were granted in the USA for the environment in Canada, Japan and the USA in 1997; for food and/or feed in the USA in 1996 and for food and use as feed in Canada in 1997. Regulatory approvals for MON809 (*goxv247*, *cry1Ab* and *CP4 epsps*) for the environment were granted in Canada and the USA in 1996 and in Japan in 1997, for food and/or feed in the USA in 1996, in Canada for food in 1996 and for use as feed in Canada in 1996 and in

Japan in 1998. Regulatory approvals for MON810 (*cry1Ab*) were granted for the environment in the USA in 1995, in Japan in 1996, in Canada and South Africa in 1997, in Argentina and the EU in 1998, for food in Canada, Japan and South Africa in 1997, in Argentina in 1998, in Australia and Switzerland in 2000 and in Korea, South Africa and Taiwan in 2002, and for use as feed in Canada, Japan and South Africa in 1997, in Argentina in 1998, in Switzerland in 2000, in South Africa in 2002 and in Mexico in 2004. It was notified as an existing product in the EU on 12 July 2004. Regulatory approvals for MON810 × LY038 (*cry1Ab*) were granted for the environment in Japan in 2007, for food and feed in the Philippines in 2006 and in Japan in 2006. Regulatory approvals for MON810 × MON88017 (*CP4 epsps*, *cry3Bb1* and *cry1Ab*) in the environment in Canada in 2006, food and/or feed in Mexico in 2006, for food in Japan in 2005, in Korea and the Philippines in 2006 and in Taiwan 2009 and for feed in the Philippines in 2006. Regulatory approvals for MON863 × MON810 (*cry3Bb1* and *cry1Ab*) were approved for the environment in Japan in 2004, for food and/or feed in Japan in 2004 and Mexico in 2006, and for food in Korea in 2009.

1.3.2 *Bacillus thuringiensis cryIF* endotoxin gene

Introduces resistance to insects

NOMENCLATURE: Approved name: *Bacillus thuringiensis* Berliner *cryIF* endotoxin gene.

Other names: *Bt* gene; *cryIF*.

Development code(s): DAS-01507-1 (event TC1507); DAS-Ø6275-8 (DAS-06275-8); TC1507 plus NK603 (crossed), TC1507 plus DAS 59122-7 (crossed), TC1507 plus DAS 59122-7 plus NK603 (crossed), MON 88017 plus MON 89034 plus TC1507 plus DAS59122-7.

Promoter(s): TC1507 – ubiquitin 1 and CaMV 35S. MON 88017 plus MON 89034 plus TC1507 plus DAS59122-7 – *pat* – CaMV 35S, *epsps* – rice actin I promoter, *cry1A.105* – CaMV 35S 5′ untranslated leader from wheat chlorophyll *a-/b*-binding protein, *cry2Ab* – FMV-35S – promoter from Figwort Mosaic Virus *Hsp70*, *cry3Bb1* – CaMV 35S promoter with duplicated enhancer region 5′ UTR from wheat chlorophyll *a-/b*-binding protein; rice actin gene first intron, *cry34Ab1* – *Zea mays* ubiquitin gene promoter, *cry35Ab* – *Triticum aestivum* peroxidase gene root-preferred promoter and *cry1Fa2* – ubiquitin (ubi) ZM (*Zea mays*) promoter.

Trait introduction: The introduction of the *cry1F* gene into DAS-Ø6275-8 (DAS-06275-8) was by *Agrobacterium tumefaciens*-mediated plant transformation. Introduction into 'Herculex' was by microparticle bombardment of plant cells or tissue. 'Herculex Xtra' is the result of a conventional breeding cross between 'Herculex I' and 'Herculex RW'. Stacked insect-resistant and herbicide-tolerant maize MON 88017 plus MON 89034 plus TC1507 plus DAS59122-7 ('Smartstax' and 'Genuity') was produced by conventional cross-breeding of parental lines MON89034, TC1507, MON88017 and DAS-59122.

SOURCE: The *cry1F* gene, isolated from the common soil bacterium *Bacillus thuringiensis* (Bt) subsp. *aizawai*, produces the insect-control protein Cry1F, a delta-endotoxin. Cry proteins, of which Cry1F is only one, act by selectively binding to specific sites localised on the lining of the midgut of susceptible insect species. Following binding, pores are formed that disrupt midgut ion flow, causing gut paralysis and eventual death due to bacterial sepsis. Cry1F is lethal only when eaten by the larvae of lepidopteran insects, and its specificity of action is directly attributable to the presence of specific binding sites in the target insects. There are no binding sites for the delta-endotoxins of *B. thuringiensis* on the surface of mammalian intestinal cells, therefore livestock animals and humans are not susceptible to these proteins.

PRODUCTION: The gene was isolated from *Bacillus thuringiensis* subsp. *aizawai*, truncated and introduced into the crop in association with a promoter, usually cauliflower mosaic virus 35S promoter, and a *pat* gene included as a selectable marker.

TARGET PESTS: Cry1F protein is effective against lepidopteran pests, and the primary target pest is the European cornborer (*Ostrinia nubilalis* Hübner). Other target pests are corn earworm (*Helicoverpa zea* Boddie), southwestern cornborer (*Diatraea grandiosella* (Dyar)), fall armyworm (*Spodoptera frugiperda* (JE Smith)), western bean cutworm (*Richia albicosta* (Smith)), black cutworm (*Agrotis ipsilon* (Hufnagel)) and pink cornborer (*Sesamia cretica* Lederer). In cotton, the target pests include tobacco budworm (*Heliothis virescens* (Fabricius)) and pink bollworm (*Pectinophora gossypiella* (Saunders)).

TARGET CROPS: Maize (corn) and cotton (plus *cryIA(c)* gene).

BIOLOGICAL ACTIVITY: The gene is expressed constitutively throughout the plant such that susceptible insects that feed upon the crop succumb. **Biology:** The CryIF endotoxin binds to a different site within the insect to CryIA and CryIB endotoxins, and it is anticipated that GM-maize containing this gene will slow the onset of insect resistance to the toxin, in addition to extending the range of insect pests controlled to include cutworms (*Agrotis* spp.) and armyworms (*Spodoptera* spp.).
Mode of action: Once in the insect, the crystal proteins are solubilised and the insect gut proteases convert the original pro-toxin into a smaller toxin. These hydrolysed toxins bind to the insect's midgut at high-affinity, specific receptor-binding sites, where they interfere

with the potassium ion-dependent active amino acid symport mechanism. This disruption causes the formation of a large cation-selective pore that increases the water permeability of the membrane. A large uptake of water causes epithelial cell swelling and eventual rupture, with the subsequent disintegration of the midgut lining. Different toxins bind to different receptors, and this explains the selectivity of different *Bt* strains. (This differential binding is described by some of those working in the field as different modes of action.). Different genes coding for these protein toxins have been isolated and used to transform various crop plants such that the crops produce insecticidal proteins throughout their life. In some cases, the truncated genes have been used so that the insect-active toxin is produced in the crop rather than the pro-toxin that requires hydrolysis in the insect gut in leaves, pollen, tassel, silk and kernel full season. **Key reference(s):** 1) RA Herman, AM Phillips, RA Collins, LA Tagliani, FA Claussen, CD Graham, BL Bickers, TA Harris & LM Prochaska. 2004. Compositional equivalency of Cry1F corn event TC6275 and conventional corn (*Zea mays* L.). *J. Agric. Food Chem.*, **52(9)**, 2726-34. 2) GD Buntin. 2008. Corn expressing Cry1Ab or Cry1F endotoxin for fall armyworm and corn earworm (Lepidoptera: Noctuidae) management in field corn for grain production. *Florida Entomologist*, **91(4)**, 523-30. 3) H Eichenseer, R Strohbehn and J Burks. 2008. Frequency and severity of western bean cutworm (Lepidoptera: Noctuidae) ear damage in transgenic corn hybrids expressing different *Bacillus thuringiensis* cry toxins. *J. Econ. Entomol.*, **101**(2), 555-63. 4) WA Siebert, JM Babock, S Nolting, AC Santos, JJ Adamczyk, PA Neese, JE King, JN Jenkins, J McCarty, GM Lorenz, DD Fromme & RB Lassiter. 2008. Efficacy of Cry1F insecticidal protein in maize and cotton for control of fall armyworm (Lepidoptera: Noctuidae). *Florida Entomologist* **91**(4): 555-65. 5) LM Appenzeller, L Malley, SA Mackenzie, D Hoban & B Delaney. 2009. Subchronic feeding study with genetically modified stacked trait lepidopteran and coleopteran resistant (DAS-Ø15Ø7-1 × DAS-59122-7) maize grain in Sprague-Dawley rats. *Food Chem Toxicol.*, **47**(7), 1512-20.

COMMERCIALISATION: Formulation: Maize containing the *cryIF* gene was commercialised by Dow AgroSciences, Mycogen and DuPont under the tradename 'Herculex 1'. It also contains the *phosphinothricin acetyl transferase* gene. Dow AgroSciences and Mycogen are evaluating the *cryIF* gene in combination with the *cryIAc* gene and the *phosphinothricin acetyl transferase* gene in cotton under the tradename 'WideStrike'. **Trade names:** 'Herculex 1' (*cry1F* plus *pat*) (event TC1507) (Dow AgroSciences), (Mycogen) and (Pioneer), 'Herculex 1 plus Roundup Ready Corn 2' (*cry1F* plus *pat* plus *epsps*) (TC1507 plus NK603 (crossed)) (Dow AgroSciences), (Mycogen), (Monsanto) and (Pioneer), 'Herculex Xtra' (*cry34A(b)1* plus *cry35A(b)1* plus *cry1F*) (event TC1507 plus DAS 59122-7 (crossed)) (Dow AgroSciences), (Pioneer) and (DuPont), 'Herculex Xtra with Roundup Ready Corn 2' (*cry34A(b)1* plus *cry35A(b)1* plus *cry1F* plus *pat* plus *epsps*) (event TC1507 plus DAS 59122-7 plus NK603 (crossed)) (Dow AgroSciences), (Monsanto), (Pioneer) and (DuPont), 'WideStrike' (*cry1A(c)* plus *cry1F* plus *pat*) (Mycogen) and (Dow

AgroSciences), 'Optimum AcreMax RW' (seed blend of 90% Herculex RW and 10% non-Bt maize) (Pioneer) and (DuPont), 'Genuity' and 'SmartStax' (*cry1A.105* plus *cry2A(b2)* plus *cry1F* plus *cry3B(b1)* plus *cry34A(b1)* plus *cry35A(b1)* plus *pat* plus *epsps*) (event MON 88017 plus MON 89034 plus TC1507 plus DAS59122-7) (Monsanto) and (Dow AgroSciences).

COMPATIBILITY: As with all GM-crops, maize containing the *crylF* gene is compatible with all the usual crop protection chemicals used in commercial maize production. DuPont has been granted approval from the US Environmental Protection Agency (EPA) for commercial registration of 'Optimum AcreMax 1' insect protection for Pioneer brand maize hybrids. This is the EPA's first approval of an in-the-bag solution for insect refuge management, offering reduced corn rootworm refuge and a more convenient path to refuge compliance for growers as refuge-in-the-bag plantings do not require the provision of untreated refugia. 'Optimum AcreMax 1' insect protection reduces the traditional 20% corn rootworm refuge by half by putting it in the seed bag, eliminating the need for separate rootworm refuge whilst increasing the ease and flexibility of planting the cornborer refuge. 'Optimum AcreMax 1' products integrate 90% of the Pioneer brand hybrid containing the 'Herculex XTRA' traits, which deliver above- and below-ground insect protection and 10% of a hybrid from the same genetic family containing the 'Herculex I' trait for above-ground insect protection and to serve as the corn rootworm refuge. All seed in the bag is glyphosate-tolerant. In addition to the 'Optimum AcreMax 1' product registration announced on 30 April 2010, the EPA also has granted Pioneer registration for 'Optimum AcreMax RW' products, which integrate 90% 'Herculex RW' seed and 10% of a hybrid from the same genetic family without biotech insect protection. 'Optimum AcreMax RW' products may be grown alone or in conjunction with 'Optimum AcreMax 1' products as a cornborer refuge, allowing growers in high corn rootworm pressure areas the flexibility to protect 100% of their maize acres with the 'Herculex RW' rootworm trait. The next-generation 'Optimum AcreMax 2' family of products would further maximise productivity and simplify insect refuge deployment through a refuge-in-the-bag solution for both corn rootworm and cornborer. 'Optimum AcreMax 2' insect protection products are anticipated for US cultivation as early as 2012 upon regulatory approval and field testing.

MAMMALIAN TOXICITY: It is considered unlikely that there will be any adverse effects of transgenic maize containing the *crylF* gene on humans. There was no significant toxicity to mice from acute oral testing of the proteins at the maximum hazard dose (Cry1F >600 mg a.i./kg and Cry1A(c) >700 mg a.i./kg).

ENVIRONMENTAL IMPACT AND NON-TARGET TOXICITY: It is expected that, as for other *Bt*-containing maize, there will be no adverse effects on non-target organisms or on the environment. **Fish Toxicity:** LC_{50} (8 days) for rainbow trout >100 mg/kg diet. No mortality or sublethal effects were observed. **Other aquatic toxicity:** There were no observed adverse effects with Cry1F and Cry1A(c) in combination at respective concentrations of 510 and 2500 μg/litre *Daphnia magna* (48 h).

Effects on beneficial insects: There were no effects on survival of honeybee larvae at 1.98 μg Cry1F + 11.94 μg Cry1A(c) per ml sugar water. The LC_{50} is >4× the pollen expression level, showing that there is no hazard to honeybee larvae or adult bees.
Behaviour in soil: The soil half-life of the plant-expressed Cry1F and Cry1A(c) was estimated as 1.3 days in a laboratory study with a representative soil from corn- and cotton-growing regions. The Cry proteins were not detectable after 14 days. These results confirm that the Cry1F and Cry1A(c) proteins degrade rapidly in soil.
DATE(S) OF REGISTRATION: Maize: Regulatory approval for 'Herculex 1' was gained for the environment and for use as food and/or feed in the USA in 2001, in Argentina in 2005 and in Brazil in 2008; food and/or feed approvals were granted in South Africa in 2002, Mexico in 2003, China in 2004 and El Salvador in 2009; food use approvals were granted in Canada and Japan in 2002, in Australia, the Philippines and Taiwan in 2003 and in Colombia and the EU in 2006; feed use approvals were granted in Canada and Japan in 2002, in the Philippines and Taiwan in 2003, in Korea in 2004 and in Colombia and the EU in 2005. Registration for 'Herculex Xtra' (event TC1507 plus DAS 59122-7 (crossed)) was granted in the USA in October 2005. Regulatory approvals for the environment were granted in Canada and Japan in 2006; for food and/or feed in Mexico in 2006; for food use in Japan in 2005, in Korea in 2006 and in the Philippines in 2007; feed use approvals were granted in Japan in 2006 and the Philippines in 2007. Regulatory approvals were granted for 'Smartstax' and 'Genuity' (Event MON 88017 plus MON 89034 plus TC1507 plus DAS59122-7) for environmental impact in Canada, Japan and the USA in 2009, for use as food and/or feed in Mexico and the Philippines in 2010 and for use in food and feed in Japan, Korea, and Taiwan in 2009.

1.3.3 *Bacillus thuringiensis cryII* endotoxin gene

Introduces resistance to insects

NOMENCLATURE:Approved name: *Bacillus thuringiensis* Berliner *cryII* endotoxin gene.

Other names: *Bt* gene; *cryII* gene; *cry2* gene.

Development code(s): LLCotton25 × MON15985 (ACS-GHØØ1-3 × MON-15985-7); MON-15985-7 × MON-Ø1445-2; MON 89034 × 1507 × MON 88017 × 59122.

Promoter(s): *cry1Ac* – double-enhanced CaMV 35S NULL; *cry2Ab* – double-enhanced CaMV 35S PetHSP70 leader sequence and the chloroplast transit peptide leader sequence; *bar* – CaMV 35S.

Trait introduction: Event 15985 ('Bollgard II') was derived from the hybrid cotton variety DP50B, which was a cross between DP50 and transgenic 'Bollgard' cotton line 531, by biolistic transformation with plasmid DNA containing the *cry2Ab* gene originally isolated from *Bacillus thuringiensis* subsp. *kurstaki*. As a result, event 15985 expresses both the Cry1Ac and Cry2Ab insecticidal proteins. MON-15985-7 × MON-Ø1445-2 is a stacked insect-resistant and herbicide-tolerant cotton derived from conventional cross-breeding of the parental lines 15985 (OECD identifier: MON-15985-7) and MON1445 (OECD identifier: MON-Ø1445-2). MON 89034 × 1507 × MON 88017 × 59122 ('Genuity' and 'SmartStax') was produced by crossing plants containing MON 89034, 1507, MON 88017 and 59122 using conventional breeding methods. MON15985 × MON88913 is a stacked insect-resistant and glyphosate-tolerant cotton produced by conventional cross-breeding of the parental lines MON88913 (OECD identifier: MON-88913-8) and 15985 (OECD identifier: MON-15985-7). Glyphosate tolerance is derived from MON88913, which contains two genes encoding the enzyme 5-enolypyruvylshikimate-3-phosphate synthase (EPSPS) from the CP4 strain of *Agrobacterium tumefaciens*. Insect resistance is derived from MON15985, which was produced by transformation of the DP50B parent variety, which contained event 531 (expressing Cry1Ac protein), with purified plasmid DNA containing the *cry2Ab* gene from *B. thuringiensis* subsp. *kurstaki*.

SOURCE: Isolated from *Bacillus thuringiensis* Berliner subsp. *kurstaki*. The *cry1A.105* gene is a chimeric gene comprising four domains from other *cry* genes previously used in transgenic plants. The amino acid sequences of Domains I and II are identical with the respective domains from Cry1Ab and Cry1Ac proteins, Domain III is almost identical to the Cry1F protein, and the C-terminal Domain is identical to Cry1Ac protein.

PRODUCTION: Monsanto transformed a 'Bollgard' cotton variety with vector GHBK11L using particle bombardment to add the *cryIIA(b)* gene. The plants are produced by either insertion of the *Bt* nucleic acid using transformed and disabled *Agrobacterium tumefaciens*, or bombardment using a particle gun or other accepted transformation techniques. Transgenic crops are identified using a selectable marker such as a herbicide-tolerance or an antibiotic-resistance gene. Elite varieties are bred from these transgenic crops.

TARGET PESTS: Lepidopteran pests, including European cornborer (*Ostrinia nubilalis* Hübner), pink cornborer (*Sesamia cretica* Lederer), noctuids (*Heliothis* spp. and *Helicoverpa* spp.), fall armyworm (*Spodoptera frugiperda* J. E. Smith) and black cutworm (*Agrotis ipsilon* (Hufnagel)). The stacked cotton line LLCotton25 × MON15985 expresses three novel proteins: the enzyme phosphinothricin acetyl-transferase (PAT) and the delta-endotoxins Cry1Ac and Cry2Ab, both of which confer resistance to the lepidopteran pests of cotton, such as the cotton bollworm (*Helicoverpa zeae* (Boddie)), pink bollworm (*Pectinophora gossypiella* (Saunders)) and tobacco budworm (*Heliothis virescens* (Fabricius)).

TARGET CROPS: Cotton and maize (corn).

BIOLOGICAL ACTIVITY: The CryII endotoxins bind to a different site within the insect to CryIA, CryIB and CryIIIB endotoxins, and it is anticipated that GM-cotton containing both genes will have improved characteristics with regard to the onset of insect resistance to the toxin in addition to extending the range of insect pests controlled. **Biology:** GM-cotton containing *cryIIA* and *cryIA* genes stacked has been commercialised under the tradename 'Bollgard II'. 'Bollgard II' is insect-resistant cotton derived by transformation of the DP50B parent variety, which contained event 531 (expressing Cry1Ac protein), with purified plasmid DNA containing the *cry2Ab* gene from *B. thuringiensis* subsp. *kurstaki*. GM-corn containing *cryIIA* and *cryIIIB* genes stacked has been commercialised under the tradename 'YieldGard Plus'.
Mode of action: See Bt *cry1A* – section 1.3.1.

COMMERCIALISATION: Trade names: 'BollGard II' (cotton – *cry1A(c)* plus *cryIIA(b)* event 15985) (Monsanto), 'YieldGard Plus' (corn – *cryIIA* plus *cryIIB*) (Monsanto), 'MON 89034' (*cry1A.105* plus *cry2A(b2)*) (Monsanto), 'Genuity VT' and 'Double PRO' (*cry1A.105* plus *cry2A(b2)* plus *epsps* – event MON 89034 NK603) (Monsanto), 'Genuity VT Triple PRO' (*cry1A.105* plus *cry2A(b2)* plus *cry3B(b1)* plus *epsps* – event MON 89034 MON 88017), 'Genuity' and 'SmartStax' (*cry1A.105* plus *cry2A(b2)* plus *cry1F* plus *cry3B(b1)* plus *cry34A(b1)* plus *cry35A(b1)* plus *pat* plus *epsps*) (event MON 88017 plus MON 89034 plus TC1507 plus DAS59122-7) (Monsanto) and (Dow AgroSciences).

COMPATIBILITY: As with all GM-crops, maize containing the *cryIIA* gene is compatible with all the usual crop protection chemicals used in commercial maize production.

MAMMALIAN TOXICITY: Acute oral toxicity: Male and female mice (10 of each) were dosed with 67, 359 and 1450 mg/kg body weight of CryIIA protein. Outward clinical signs were observed and body weights recorded throughout the 14-day study. Gross necropsies performed at the end of the study indicated no findings of toxicity attributed to exposure to the test substance. No mortality or clinical signs attributed to the test substance were noted during the study. Tests have demonstrated that the CryIIA delta-endotoxin is rapidly degraded by gastric fluid *in vitro*. In a solution of simulated gastric fluid (SGF) (pH 1.2), complete degradation of detectable CryIIA protein occurred within 15 s. Incubation in SGF resulted in a ~50 kDa protein digestion product. A comparison of amino acid sequences of known allergens uncovered no evidence of any homology with CryIIA, even at the level of eight contiguous amino acids residues. Thus, the potential for the CryIIA protein to be a food allergen is minimal.

ENVIRONMENTAL IMPACT AND NON-TARGET TOXICITY: No significant or irreversible hazards from CryIIA maize or cotton to non-target organisms or to the environment are anticipated. For larval honeybees, the no observed effects concentration (NOEC) was found to be greater than 100 ppm, which was expected to exceed the maximum concentration encountered in the field. Adult honeybee NOEC was greater than 68 ppm. Green lacewing larval NOEC was greater than 1100 ppm. All these exceeded the maximum environmental concentration expected in cotton plant tissue (51 ppm). The LC_{50} of Cry2Ab2 protein for ladybirds was >4500 ppm, which greatly exceeded the level expected in the field.

DATE(S) OF REGISTRATION: 'Bollgard II' (*cry1A*(*c*) plus *cryIIA*(*b*)) received regulatory approval for the environment in Australia and the USA and for food and/or feed use in the USA and for food use in Australia in 2002; Japan gave approval for food use in 2002 and for feed use in 2003; in 2003 it was granted approval for the environment and for food and/or feed use in Australia, and for food and/or feed use in Canada and the Philippines; Korea gave approval for use as food in 2003 and feed in 2004; India gave environmental approval and China food and/or feed approval in 2006; Burkina Faso gave environmental approval in 2008; and Brazil granted environmental and food and/or feed approval in 2009. MON15985 (*cry1A*(*c*) plus *cry2A*(*b*) genes) was placed on the market in the USA on 1 January 2003, but applications have been withdrawn. The European Food Safety Authority (EFSA) received the applications on 6 January 2005 and they were accepted on 16 September 2005. However, additional information has been requested and the proceedings have been halted. An application for renewal of an existing product according to Regulation 1829/2003 on genetically modified food and feed was requested for MON 15985 101 × MON 1445 (*cryIA*(*c*) plus *cry2A*(*b2*) plus *cp4 epsps* genes). This was checked for completeness and accepted on 9 April 2008 by EFSA. However, additional information was requested from the applicant and the proceedings have been suspended. An application for MON15985 × MON88913 (*cryIa*(*c*) plus *cry2a*(*b2*) plus *cp4 epsps* genes) was made in April 2007 and checked for completeness by the EFSA. Additional information has been requested from the applicant and the proceedings have been suspended. Environmental impact applications were approved for MON89034 maize (*cry1A.105* plus *cry2Ab*) in the USA, Japan and Canada in 2008 and in Brazil in 2009. It was approved for food and/or feed in the USA in 2007 and in the EU and Brazil in 2009. It was cleared for food use in Japan in 2007, in Australia, Canada, and Taiwan in 2008 and in the Philippines in 2009. It was approved for use as feed in Canada, Colombia and Japan in 2008 and in Korea and the Philippines in 2009. Food and/or feed approvals for MON89034 × MON88017 (*cry1A.105* plus *cry2Ab*(*2*) plus *cry3Bb*(*1*) plus *CP4 epsps* genes) were granted in Japan in 2008, in the Philippines in 2009 and in Mexico in 2010. It was approved for food use in Korea and Taiwan in 2009 and for use as feed in Taiwan in 2009. Regulatory approvals for MON89034 × NK603 (*cry1A.105* plus *cry2Ab* plus *CP4 epsps* (rice actin 1 promoter) plus *CP4 epsps* (enhanced CaMV 35S)-

genes) were granted for food and/or feed in Japan in 2008 and in Mexico in 2010, for food in the Philippines and Taiwan in 2009 and in Korea in 2010, and for use as feed in Korea and the Philippines in 2009. Regulatory approvals for MON89034 × TC1507 × MON88017 × DAS-59122-7 (*pat* plus *CP4 epsps* plus *cry1A.105* plus *cry2Ab* plus *cry3Bb1* plus *cry34Ab1* plus *cry35Ab1* plus *cry1Fa2*) were granted for environmental impact in Canada, Japan and the USA in 2009, for food and/or feed in Mexico and the Philippines in 2010, and for food and use as feed in Japan, Korea and Taiwan in 2009. Regulatory approvals for 15985 × MON1445 (*CP4 epsps* plus *cry1Ac* plus *cry2Ab*) were achieved in Australia for the environment in 2002; in the Philippines for food and feed use in 2004; in Korea for food use in 2004 and feed use in 2008; in the EU and Japan for food and/or feed use in 2005; and in Mexico for food and/or feed use in 2006. MON-15985-7 × MON-88913-8 (*CP4 epsps* plus *cry1Ac* plus *cry2Ab*) received regulatory approval for use as feed in 2005 and for use in food in 2006 in Japan; environmental approval was granted in Australia in 2006 and South Africa in 2007; food and/or feed approval was granted in Mexico in 2006 and South Africa in 2007; food use was approved in Korea and the Philippines in 2006, and feed use was approved in the Philippines in 2006 and Korea in 2008. MON 89034 was registered for the environment in Canada, Japan and the USA in 2008 and in Brazil in 2009; it received approval for food and/or feed use in the USA in 2007 and in Brazil and Colombia in 2009; food use clearance was granted in Japan in 2007, in Australia, Canada and Taiwan in 2008 and in Korea and the Philippines in 2009; feed clearance was granted in Canada, Colombia, Japan and Korea in 2008 and in Korea and the Philippines in 2009.

1.3.4 *Bacillus thuringiensis cryIIIA* endotoxin gene

Introduces resistance to insects

NOMENCLATURE: Approved name: *Bacillus thuringiensis* Berliner *cryIIIA* endotoxin gene.

Other names: *Bt* gene; *cryIIIA* gene; *cry3A* gene.

Promoter(s): Cauliflower mosaic virus (CaMV) 35S.

SOURCE: *Bacillus thuringiensis* produces parasporal, protein, crystalline inclusion bodies during sporulation.

PRODUCTION: Transgenic potatoes produced by genetic engineering using plant-expressed vectors that transferred the *cryIIIA* and the selectable marker, *neomycin phosphotransferase II* (*nptII*), genes into the genomic DNA of the potato plants. These transformants were used to produce commercial seed potatoes through conventional breeding programmes. *Events:* 'NewLeaf' BT6, BT10, BT12, BT16, BT17, BT18 and BT23. 'NewLeaf Plus' RBMT21-129 and 21-350.

TARGET PESTS: Colorado beetle (CB) (*Leptinotarsa decemlineata* (Say.)).

TARGET CROPS: Potatoes.

BIOLOGICAL ACTIVITY: The transformed potato plants express the *Bt* endotoxin throughout the plant. The CB ingests the toxin whilst feeding on the plant. The endotoxin is activated in the CB's gut by enzymes. The toxin binds to membranes of the gut epithelial cells, and pores are formed. Cells in the gut rupture and the CB larvae die. The Cry IIIA delta-endotoxin produced in potatoes is identical to that found in Nature and in commercial *Bt* formulations. However, the potatoes produce the Cry IIIA delta-endotoxin throughout the plant for the length of the growing season, at a level sufficient to control all life stages of the CB. In contrast, the application of foliar *Bt* insecticide must be frequent and carefully timed to protect the crop adequately. **Mode of action:** See Bt cry1A – section 1.3.1. The mechanism of action for the Cry3A protein produced by *Bacillus thuringiensis* subsp. *tenebrionsis* (Btt) and by 'NewLeaf' potato plants has been well defined. Ingestion of the Cry3A protein produced by 'NewLeaf' potato plants stops insect feeding, eventually leading to death by starvation. The mechanism of feeding inhibition relies on highly specific binding of the Cry3A protein to cells lining the insect gut. The ligand receptor-based binding causes a disruption in ion flow leading to pore formation in the cells, thus destroying the integrity of the gut cells. **Efficacy:** Field experiments conducted at more than 30 locations throughout the US potato-growing region since 1991 have demonstrated that *Bt* potatoes are protected season-long from all CB life stages. Growers who use *Bt*-potatoes do not require chemical insecticide applications to control CB. The Long Island, New York potato production area has CB populations that are highly resistant to most chemical insecticides. *Bt*-potatoes were tested on Long Island and provided excellent, season-long control of all stages of CB and produced high yields, without relying on chemicals for control. However, recently introduced foliar-applied insecticides have been shown to be effective against CB (including those resistant to earlier insecticides) and against other common insect pests of potatoes, such as aphids. This has led to a significant fall in the planting of transgenic potato seed and the subsequent withdrawal of the products. **Key reference(s):** FS Betz, BG Hammond & RL Fuchs. 2000. Safety and advantages of *Bacillus thuringiensis*-protected plants to control insect pests. *Regulatory Toxicology and Pharmacology*, **32**, 156-73.

COMMERCIALISATION: Formulation: 'NewLeaf' potatoes were introduced by the Monsanto subsidiary, NatureMark, in 1995, with additional virus-resistant 'NewLeaf' varieties being introduced from 1998. In 'NewLeaf', the transformed variety is Russet Burbank. **Trade names:** 'NewLeaf', 'NewLeaf Plus' (plus PLRV coat protein) and 'NewLeaf Y' (plus PVY coat protein) (NatureMark) (withdrawn).

COMPATIBILITY: 'NewLeaf' potatoes are compatible with all crop protection chemicals that are used in potato cultivation. It is unlikely that the same quantity of insecticide would be needed as in a conventional crop.

MAMMALIAN TOXICITY: The receptors for the Cry3A protein are not present on mammalian cells nor in insects not closely related to CB. Hence, the Cry3A protein is effective for control of a limited number of coleopteran insects and only those insects that might feed on potato tissue produced by 'NewLeaf' potato plants. Molecular characterisation of CB-resistant Russet Burbank potatoes showed equivalence to microbially produced *Btt* protein. The relative size and number of copies of the DNA inserted into potatoes was demonstrated with endonuclease-digested chromosomal DNA from field-grown potato plants southern blotted with the entire introduced plasmid as the probe. These southern blots provided information about the number of copies of introduced DNA, the lack of significant amount of DNA introduced outside the border regions and the integrity of the introduced DNA near the endonuclease cut site. These results indicate that only the DNA necessary to produce the CryIIIA delta-endotoxin was introduced into the plant, thus indicating that exposure would be only to the CryIIIA delta-endotoxin and the nucleic acids found in the genetic material necessary for its production. Such nucleic acids have not, by themselves, been associated with toxic effects to animals or humans and are regular constituents of the human diet. Equivalence of microbially produced and plant-produced *Btt* protein, also called Colorado beetle active protein from *Bacillus thuringiensis* subsp. *tenebrionis*, has been demonstrated. Microbially produced delta-endotoxins from the *cryIIIA* gene, as expressed in *E. coli* and in potato tubers, were compared. These comparisons included SDS-PAGE co-migration, western blot analysis, staining for carbohydrate residues, *N*-terminal amino acid sequence analysis and biological equivalence against *L. decemlineata*. These data are adequate to support the equivalence of the microbially produced and plant-produced protein for use in the toxicology studies. **Acute oral toxicity:** The *Btt* proteins were determined to be stable, and the dosing concentrations for mice were determined to be 74.9 mg/ml, 14.62 mg/ml and 7.4 mg/ml. *Btt* protein was not toxic by oral gavage when mice were dosed with up to 5220 mg/kg body weight. These results placed this protein in Toxicity Category IV.

***In vitro* digestibility of *Btt* protein:** The 68 kDa and 55 kDa *Btt* proteins degraded within 30 s in simulated gastric fluid when analysed by western blot and were not active against Colorado beetles after degradation. The 68 kDa *Btt* protein degraded to 55 kDa within 2 h of incubation in simulated intestinal fluid, while the 55 kDa form remained unchanged after 14 h of incubation and retained its bioactivity and western blot results. These results indicate that, following ingestion by humans, the *Btt* proteins will be degraded like other proteins to amino acids and peptides similar to those occurring in a normal human diet. It was concluded by the regulatory authorities that *Btt*-potatoes present little potential for human dietary toxicity. At a dose one million-fold greater than that contained in a potato, no toxicity was observed. Moreover, several studies showed that *Btt*-potatoes are indistinguishable from strains of conventional potatoes in nutritive content (total protein, total sugars, vitamin C, minerals). Furthermore, the *Btt* toxin is rapidly digested by pepsin and is inactivated by heat encountered in cooking.

ENVIRONMENTAL IMPACT AND NON-TARGET TOXICITY: Experimental field trials and use on commercial farms have shown that the Colorado beetle (CB)-resistant Russet Burbank potatoes: 1) exhibit no plant-pathogenic properties; 2) are no more likely to become weeds than non-transgenic Russet Burbank or CB-resistant potatoes that could potentially be developed by traditional breeding techniques; 3) are unlikely to increase the weediness potential of any other cultivated plant or native wild species with which the organisms can interbreed; 4) should not cause damage to raw or processed agricultural commodities; 5) are unlikely to harm other organisms, such as bees, which are beneficial to agriculture, or to adversely impact the ability to control non-target insect pests; and 6) should pose no greater threat to the ability to control CB in potatoes and other crops, other than that posed by the widely practised method of applying insecticides to control CB on potatoes.

DATE(S) OF REGISTRATION: 'NewLeaf' varieties were registered in the USA in 1995 and in Canada in 1996. 'NewLeaf Plus' was registered in 1998 in the USA and in 1999 in Canada. 'NewLeaf Atlantic' and 'NewLeaf Superior' were registered in the USA in 1999 and in Canada in 2000. A change in emphasis by Monsanto has led to all 'NewLeaf' varieties being withdrawn in 2001, although the US and Canadian registrations are still valid.

1.3.5 *Bacillus thuringiensis cryIIIB* endotoxin gene

Introduces resistance to insects

NOMENCLATURE: Approved name: *Bacillus thuringiensis* Berliner *cryIIIB* endotoxin gene.

Other names: *Bt* gene; *cryIIIB* gene; *cry3B* gene.

Promoter(s): Cauliflower mosaic virus (CaMV) 35S.

SOURCE: *Bacillus thuringiensis* produces parasporal, protein, crystalline inclusion bodies during sporulation. Different subspecies produce different endotoxins with different spectra of activity. *B. thuringiensis* subsp. *kumamotoensis* produces Cry IIIB proteins and *B. thuringiensis* subsp. *kurstaki* produces Cry IIA proteins.

PRODUCTION: The gene coding for CryIIIB proteins was isolated from *B. thuringiensis* as part of a collaboration between Monsanto and Ecogen. A truncated version has been introduced into maize (corn), often in conjunction with a truncated *cry IIA* gene.

TARGET PESTS: Corn rootworm (*Diabrotica* spp.) and lepidopteran pests, including European cornborer (*Ostrinia nubilalis* Hübner), pink cornborer (*Sesamia cretica* Lederer), noctuids (*Heliothis* spp. and *Helicoverpa* spp.) and fall armyworm (*Spodoptera frugiperda* J. E. Smith).

TARGET CROPS: Maize (corn).

BIOLOGICAL ACTIVITY: Biology: The *cryIIIB* gene codes for a truncated endotoxin that is active against corn rootworm (*Diabrotica* spp.) and the *cryIIA* gene codes for the lepidopteran-specific endotoxin.
Mode of action: See Bt cry1A – section 1.3.1.
Efficacy: Preliminary field studies have shown good control of the target insects in the absence of additional chemical insecticide applications. The stacking of two genes with different activity spectra ensures that whatever insect pests appear in the crop, additional chemical treatments are unnecessary. **Key reference(s):** J Payne, J Fernandez-Cornejo & S Daberkow. 2003. Factors affecting the likelihood of corn rootworm Bt seed adoption, *AgBioForum*, **6(1&2)**, 79-86. (Available on the World Wide Web: www.agbioforum.org).

COMMERCIALISATION: Trade names: 'YieldGard Plus' (Monsanto).

COMPATIBILITY: As with all GM-crops, maize containing the *cryIIA* and *cryIIIB* gene is compatible with all the usual crop protection chemicals used in commercial maize production.

MAMMALIAN TOXICITY: Acute oral toxicity: Male and female mice (10 of each) were dosed with 67, 359 and 1450 mg/kg body weight of CryIIA protein. Outward clinical signs were observed and body weights recorded throughout the 14-day study. Gross necropsies performed at the end of the study indicated no findings of toxicity attributed to exposure to the test substance. No mortality or clinical signs attributed to the test substance were noted during the study. Tests have demonstrated that the CryIIA delta-endotoxin is rapidly degraded by gastric fluid *in vitro*. In a solution of simulated gastric fluid (SGF) (pH 1.2), complete degradation of detectable CryIIA protein occurred within 15 s. Incubation in SGF resulted in a ~50 kDa protein digestion product. A comparison of amino acid sequences of known allergens uncovered no evidence of any homology with CryIIA,

1.3 Insect resistance

even at the level of eight contiguous amino acids residues. Thus, the potential for the CryIIA protein to be a food allergen is minimal. Two acute oral studies have been undertaken on CryIIIB(b)1 proteins. These studies were done with two variants of the CryIIIB(b)1 protein engineered with either four or five internal amino acid sequence changes to enhance activity against the corn rootworm. Acute oral toxicity: data submitted support the prediction that the CryIIIB(b)1 protein would be non-toxic to humans. Male and female mice were dosed with 36, 396, or 3780 mg/kg body weight of CryIIIB(b)1 protein for one variant. The mice were dosed with 38.7, 419, or 2980 mg/kg body weight of CryIIIB(b)1 protein for the other variant. In one study, two animals in the high-dose group died within a day of dosing. These animals both had signs of trauma, probably due to dose administration (lung perforation or severe discoloration of lung, stomach, brain and small intestine). No clinical signs were observed in the surviving animals and normal body weight gains were recorded throughout the 14-day study for the remaining animals. Gross necropsies performed at the end of the study indicated no findings of toxicity attributed to exposure to the test substance in either study. No other mortality or clinical signs attributed to the test substance were noted during either study. It has been demonstrated that the CryIIIB(b)1 protein is rapidly degraded by gastric fluid *in vitro*. In a solution of simulated gastric fluid (SGF) (pH 1.2), complete degradation of detectable CryIIIB(b)1 protein occurred within 30 s. Insect bioassay data indicate that the protein loses insecticidal activity within 2 min of incubation in SGF. Incubation in simulated intestinal fluid resulted in a *ca.* 59 kDa protein digestion product. A comparison of amino acid sequences of known allergens uncovered no evidence of any homology with CryIIIB(b)1, even at the level of eight contiguous amino acid residues.

ENVIRONMENTAL IMPACT AND NON-TARGET TOXICITY: No significant or irreversible hazards from maize producing CryIIA plus CryIIIB(b) to non-target organisms or to the environment are anticipated.

1.3.6 *Bacillus thuringiensis* subspp. *kumamotoensis cry3A* and *B* endotoxin gene

Introduces resistance to insects

NOMENCLATURE: Approved name: *Bacillus thuringiensis* subspp. *kumamotoensis cryIIIB(bl)* endotoxin gene.

Other names: *Bt* gene; *cryIIIB(b)* gene; *cry3B(b)* gene, *cryIIIA(b)* gene; *cry3A(b)* gene.

Development code(s): MON 863; MON810 × MON88017.

Promoter(s): *cry3B(b)* – CaMV 35S promoter with duplicated enhancer region 5′ UTR from wheat chlorophyll *a-/b*-binding protein; rice actin gene first intron.

Trait introduction: MON 863 was developed by microparticle bombardment of plant cells or tissue. MON810 × MON88017 (OECD identifier: MON-ØØ81Ø-6 × MON-88Ø17-3) maize is an F1 hybrid resulting from the hybridisation of maize-inbred MON 810 (MON-ØØ81Ø-6) with MON88017 (MON-88Ø17-3). This stacked maize hybrid is a product of traditional plant breeding.

SOURCE: CryIIIB(bl) protein is a delta-endotoxin from *Bacillus thuringiensis* subspp. *kumamotoensis* and has activity against certain beetles.

PRODUCTION: The wild-type *cryIIIB(bl)* gene was modified to enhance the protein's activity against the corn rootworm complex (*Diabrotica* spp.). Two CryIIIB(bl) variants were engineered for expression in the bacterium *Bacillus thuringiensis* isolates EG11098 and EG11231. CryIIIB(bl) protein resulting from these isolates differed from wild-type CryIIIB(bl) protein at five and four amino acid positions, respectively. Corn was genetically modified to express the CryIIIB(bl).11231 protein (resulting in corn line MON 853) or the CryIIIB(bl).11098 protein (resulting in corn line MON863). Ranges of CryIIIB(bl) protein levels in MON863 in μg CryIIIB(bl) protein per g of fresh weight tissue were 30–93 (leaf), 49–86 (grain), 30–93 (pollen), 3.2–66 (root) and 13–54 (above-ground whole plant).

TARGET PESTS: Corn rootworm complex (*Diabrotica* spp.).

TARGET CROPS: Maize (corn).

BIOLOGICAL ACTIVITY: The transformed corn plants express the *Bt*-endotoxin throughout the plant, including the roots. Corn rootworm larvae ingest the toxin whilst feeding on the plant's roots. The endotoxin is activated in the larval gut by enzymes. The toxin binds to membranes of the gut epithelial cells, and pores are formed. Cells in the

gut rupture and the larvae die (see section 1.3.1). These GM-corn plants produce the Cry IIIBb(I) delta-endotoxin for the length of the growing season, at a level sufficient to control all life stages of the corn rootworm. In contrast, the corn rootworm larvae cannot be controlled by foliar application of an insecticide because they inhabit the soil. It is common practice to use persistent, relatively toxic insecticides to control these pests.
Key reference(s): J Payne, J Fernandez-Cornejo & S Daberkow. 2003. Factors affecting the likelihood of corn rootworm Bt seed adoption, *AgBioForum*, **6(1&2)**, 79-86. (Available on the World Wide Web: www.agbioforum.org).

COMMERCIALISATION: Trade names: 'YieldGard Rootworm' (*cry3B1b*) (Monsanto), 'Agrisure 3000GT' (*cry1A(b)* plus *cry3A* plus *pat* plus *epsps* – event BT11 MIR604) (Syngenta), 'Genuity VT Triple PRO' (*cry1A.105* plus *cry2A(b2)* plus *cry3B(b1)* plus *epsps* – event MON 89034 MON 88017), 'Genuity' and 'SmartStax' (*cry1A.105* plus *cry2A(b2)* plus *cry1F* plus *cry3B(b1)* plus *cry34A(b1)* plus *cry35A(b1)* plus *pat* plus *epsps*) (Monsanto) and (Dow AgroSciences).

MAMMALIAN TOXICITY: Mice dosed with rates up to 2700 mg/kg b.w. of Cry3Bb1 protein showed no findings of toxicity attributed to exposure to the test substance, and no mortality or clinical signs attributed to the test substance were noted. Current scientific knowledge suggests that common food allergens tend to be resistant to degradation by heat, acid and proteases, and may be glycosylated and present at high concentrations in the food. Data have been submitted that demonstrate that the Cry3Bb1 protein is rapidly degraded by gastric fluid *in vitro*. In a solution of simulated gastric fluid (SGF) (pH 1.2), complete degradation of detectable Cry3Bb1 protein occurred within 30 s. Insect bioassay data indicated that the protein lost insecticidal activity within 2 min of incubation in SGF. The Bt protein expressed in maize MON 863 (Cry3Bb1) was shown to be equivalent to microbial Bt used in sprays for over 30 years. The history of use and literature suggest that the bacterial Bt protein is not toxic to humans, other vertebrates and non-coleopteran invertebrates. This protein is active only against specific coleopteran insects. There are no coleopteran species listed as threatened or endangered in Canada or the USA.

ENVIRONMENTAL IMPACT AND NON-TARGET TOXICITY: Environmental impact data have demonstrated that there is virtually no possibility of any risk associated with weediness or outcrossing to wild relatives. Field studies indicated that MON863 did not have a negative effect on the abundance of non-target organisms compared with a non-Bt hybrid. Species studied included arthropods, earthworms and soil microbes. Coleopteran families including Carabidae, Staphylinidae and Coccinellidae were observed. Consideration was also given to the exposure of fireflies (family Lampyridae) to MON 863, since fireflies are coleopterans. It was concluded that the possibility of firefly exposure to Cry3Bb1 protein from MON 863 would be very unlikely, as these insects generally do not occur sub-surface and do not feed on maize. Feeding trials on non-target invertebrates, including honeybees, ladybird beetles, daphnia, collembola and earthworms, were also conducted. In addition, feeding trials were conducted on non-target vertebrates such catfish and quail. MON 863 was demonstrated to be safe to these indicator species. MON 863 also did not demonstrate an increased level of anti-nutritional factors such as trypsin inhibitor and phytic acid when compared with unmodified control lines.

DATE(S) OF REGISTRATION: 'YieldGard Rootworm' was registered for food and/or feed in the USA in 2001, in Mexico in 2003 and in China in 2004; it was approved for use in food in Japan in 2002, in Australia, Canada, Korea, the Philippines and Taiwan and in the EU in 2006; feed use was approved in Japan in 2002, in Canada and the Philippines in 2003, in Korea in 2004 and in the EU in 2005. Food and/or feed approvals for MON89034 × MON88017 (*cry1A.105*, *cry2Ab*(2), *cry3Bb*(1) and *CP4 epsps*-genes) were granted in Japan in 2008, in the Philippines in 2009 and in Mexico in 2010. It was approved for food use in Korea and Taiwan in 2009 and for use as feed in Taiwan in 2009. Regulatory approvals for MON89034 × TC1507 × MON88017 × DAS-59122-7 (*pat*, *CP4 epsps*, *cry1A.105*, *cry2Ab*, *cry3Bb1*, *cry34Ab1*, *cry35Ab1* and *cry1Fa2*) were granted for environmental impact in Canada, Japan and the USA in 2009, for food and/or feed in Mexico and the Philippines in 2010, and for food and use as feed in Japan, Korea and Taiwan in 2009. Regulatory approvals were granted for MON810 × MON88017 (*CP4 epsps*, *cry3Bb1* and *cry1Ab*) in the environment in Canada in 2006, food and/or feed in Mexico in 2006, for food in Japan in 2005, in Korea and the Philippines in 2006 and in Taiwan 2009 and for feed in the Philippines in 2006. Regulatory approvals for MON863 × MON810 (*cry3Bb1* plus *cry1Ab*) were approved for the environment in Japan in 2004, for food and/or feed in Japan in 2004 and Mexico in 2006, and for food in Korea in 2009.

1.3.7 *Bacillus thuringiensis cry9C* endotoxin gene

Introduces resistance to insects

NOMENCLATURE: Approved name: *Bacillus thuringiensis* Berliner *cry9C* endotoxin gene.

Other names: *Bt* gene; *cry9C* gene.

Development code(s): Maize line CBH-351 ('StarLink').

Promoter(s): *pat* gene – cauliflower mosaic virus (CaMV) 35S; *cry9C* gene – CaMV 35S; leader sequence of the *cab22L* gene of *Petunia hybridia* (chloroplast transit peptide).

Trait introduction: Trait was introduced by microparticle bombardment of plant cells or tissue.

SOURCE: Isolated from *Bacillus thuringiensis* Berliner subsp. *tolworthi* strain BTS02618A.

PRODUCTION: The gene was isolated from *B. thuringiensis* subsp. *tolworthi* and introduced into maize (corn) in conjunction with the glufosinate-resistance gene (*pat*), included both as a selectable marker and to introduce tolerance to glufosinate-ammonium (Liberty) (event CBH351).

TARGET PESTS: European cornborer (*Ostrinia nubilalis* Hübner), pink cornborer (*Sesamia cretica* Lederer) and noctuids (*Heliothis* spp. and *Helicoverpa* spp.).

TARGET CROPS: Maize (corn).

BIOLOGICAL ACTIVITY: The gene is expressed constitutively throughout the plant such that susceptible insects that feed upon the crop succumb. Cry9C endotoxin binds at a different site to CryIA endotoxin and, therefore, has a place in insect-resistance management strategies. **Mode of action:** For a full description of the mode of action and insect specificity of *Bt*-endotoxins, see the *Bacillus thuringiensis* Berliner entries in section 1.3.1. The endotoxin showed one property that is typically found in common human allergens, and use of the gene was therefore restricted to animal feed-grain. It was subsequently found in tacos produced for human consumption and was withdrawn by the manufacturer.

COMMERCIALISATION: Formulation: Introduced by Aventis (now Bayer CropScience) as 'StarLink' corn for use only as an animal feed-grain. **Trade names:** 'StarLink' (plus glufosinate-tolerance) (withdrawn) (Aventis).

MAMMALIAN TOXICITY: Restricted clearance was given to 'StarLink' corn, as there was evidence that the Cry9C endotoxin was not broken down rapidly in simulated gastric fluid. This property is one of six US-EPA indicators to warn of potential allergenicity; results for the other five were negative. Its illegal use in grain processed for human consumption led to its withdrawal.

ENVIRONMENTAL IMPACT AND NON-TARGET TOXICITY: Field testing has demonstrated that CBH-351 maize plants are well protected from European cornborer and exhibit tolerance to glufosinate ammonium herbicides at concentrations that provide effective weed control. CBH-351 maize has undergone field testing in a wide variety of locations, in 31 states and territories of the USA from 1995 and in Canada, Belgium, France, Chile and Argentina. This field testing was conducted, in part, to confirm that CBH-351 maize exhibits the desired agronomic characteristics and to demonstrate that CBH-351 maize does not pose a plant pest risk. The kernel composition, quality and other characteristics of CBH-351 maize were found to be similar to non-transgenic maize and should have no adverse impact on raw or processed agricultural commodities. Studies were conducted on several non-target organisms to determine the potential toxic effects of Cry9C protein on test organisms, including adult honeybees (*Apis mellifera* L.), a predator ladybird beetle (*Hippodamia convergens* Guèrin), juveniles of the soil-dwelling invertebrate Collembola (springtails) (*Folsomia candida* Willem), earthworms, juveniles of the freshwater invertebrate *Daphnia magna* Straus, northern bobwhite chicks, and mice. The Cry9C protein was expressed either in whole-plant powder or pollen derived from CBH-351 maize plants, or as purified from a Cry-minus *Bt* bacterial strain engineered to express the protein toxin. Control maize plant powder and pollen test substances lacking insecticidal activity (as assayed against the European cornborer) were used in these studies to determine whether effects were specific to the CBH-351 transformation event. No effects on these organisms were detected during any of these studies that could be related to the presence of the insecticidal Cry9C protein in CBH-351 maize.

DATE(S) OF REGISTRATION: Regulatory approval was granted in the USA for the environment and for food use in 1998. Applications were considered by Argentina in 1998. The product has since been withdrawn.

1.3.8 Bacillus thuringiensis *cry34* and *cry35* endotoxin genes

Introduces resistance to insects

NOMENCLATURE: Approved name: *Bacillus thuringiensis cry34* endotoxin gene; *Bacillus thuringiensis cry35* endotoxin gene.

Other names: *Bt* gene; *cry34A* gene; *cry35A* gene.

Development code(s): DAS-59122-7 ('Herculex RW'); TC1507; DAS 59122-7; NK603; DAS-59122-7 × NK603.

Promoter(s): *cry34Ab1* – *Zea mays* ubiquitin gene promoter, *cry35Ab* – *Triticum aestivum* peroxidase gene root-preferred promoter.

Trait introduction: The introduction of the *cry34A* and *cry35A* genes into DAS-Ø6275-8 (DAS-06275-8) was by *Agrobacterium tumefaciens*-mediated plant transformation. 'Herculex Xtra' is the result of a conventional breeding cross between 'Herculex I' and 'Herculex RW'.

DAS-59122-7 maize was produced by *Agrobacterium*-mediated transformation of the hybrid maize line Hi-II.

SOURCE: The *cry34Ab1* and *cry35Ab1* genes were isolated from the common soil bacterium *Bacillus thuringiensis* Berliner (Bt) strain PS149B1.

PRODUCTION: Stacked insect-resistant and herbicide-tolerant maize MON 88017 plus MON 89034 plus TC1507 plus DAS59122-7 ('Smartstax' and 'Genuity') was produced by conventional cross-breeding of parental lines MON89034, TC1507, MON88017 and DAS-59122.

TARGET PESTS: The insect-control proteins (delta-endotoxins) Cry34Ab1 and Cry35Ab1, which are insecticidal to coleopterans such as the corn rootworm (*Diabrotica undecimpunctata* Barber).

TARGET CROPS: Maize (corn).

BIOLOGICAL ACTIVITY: The transformed corn plants express the *Bt* endotoxin throughout the plant, including the roots. Corn rootworm larvae ingest the toxin whilst feeding on the plant's roots. The endotoxin is activated in the larval gut by enzymes. The toxin binds to membranes of the gut epithelial cells, and pores are formed. Cells in the gut rupture and the larvae die. These GM-corn plants produce the Cry34A and Cry35A delta-endotoxins for the length of the growing season, at a level sufficient to control all life stages of the corn rootworm. In contrast, the corn rootworm larvae cannot be controlled by foliar application of an insecticide because they inhabit the soil. It is common practice to use persistent, relatively toxic insecticides to control these pests. **Biology:** The *cry34* and *cry35* genes code for endotoxins that are active against corn rootworm (*Diabrotica* spp.). **Mode of action:** For details of the mode of action see *Bacillus thuringiensis cry 1* gene entry 1.3.1. **Efficacy:** Field studies have shown good control of the target insects in the absence of additional chemical applications. Where other insect pests have appeared in the crop, additional chemical treatment is usually recommended.

COMMERCIALISATION: Trade names: 'Herculex RW' (*cry34A(b)1* plus *cry35A(b)1* plus *pat*) (Pioneer) and (DuPont), 'Herculex Xtra' (*cry34A(b)1* plus *cry35A(b)1* plus *cry1F*) (Pioneer) and (DuPont), 'Herculex Xtra with Roundup Ready Corn 2' (*cry34A(b)1* plus *cry35A(b)1* plus *cry1F* plus *pat* plus *epsps*) (event TC1507 plus DAS 59122-7 plus NK603 (crossed)) (Dow AgroSciences), (Monsanto), (Pioneer) and (DuPont), 'Optimum AcreMax RW' (seed blend of 90% Herculex RW and 10% non-*Bt* maize) (Pioneer) and (DuPont), 'Genuity' and 'SmartStax' (*cry1A.105* plus *cry2A(b2)* plus *cry1F* plus *cry3B(b1)* plus *cry34A(b1)* plus *cry35A(b1)* plus *pat* plus *epsps*) (Monsanto) and (Dow AgroSciences).

COMPATIBILITY: As with all GM-crops, maize containing the *cryIF* gene is compatible with all the usual crop protection chemicals used in commercial maize production.

MAMMALIAN TOXICITY: The acute oral toxicity of the two Cry proteins was studied, both individually and combined, using single-dose studies in mice. The mice received either a single dose of 2700 mg/kg b.w. Cry34Ab1, 1850 mg/kg bw Cry35Ab1 or 482 mg/kg bw Cry34Ab1 and 1520 mg/kg bw Cry35Ab1 together, and observed for 2 weeks. Apart from some fluctuations in body weight common in large-dose oral gavage studies, there were no clinical signs noted during the study. All mice survived the observation period and there were no gross pathological changes noted upon post mortem examination of the mice.

ENVIRONMENTAL IMPACT AND NON-TARGET TOXICITY: Acute dietary toxicity studies were conducted in laboratory tests on the effect of the Cry34 and Cry35Ab1 proteins on non-target invertebrates, including honeybee larvae, green lacewing larvae, ladybirds (*Hippodamia convergens* Guérin-Méneville and *Coleomegilla maculate* (De Geer)), parasitic wasps, ground beetles, daphnia, collembola, and earthworms. Data were also submitted on non-target vertebrates including the rainbow trout. Cry34 and Cry35Ab1 proteins were demonstrated to be safe to these indicator species when ingested alone or in combination, at doses exceeding the levels of direct or indirect exposure to Cry34 and Cry35Ab1 proteins from corn 59122 tissues. Although growth inhibition was reported when *C. maculate* larvae were fed Cry34Ab1 protein at about 10 times the field pollen value, no delay in development or weight reduction were observed when larvae were fed a diet composed of 50% corn line 59122 pollen, which represents a conservative estimate of the potential exposure in the field. Therefore, no sublethal effects on *C. maculate* ladybirds are expected from exposure to event DAS-59122-7.

DATE(S) OF REGISTRATION: The registration for 'Herculex Xtra' (Event TC1507 × DAS59122-7) was granted in the USA in October 2005. Regulatory approvals for MON89034 × TC1507 × MON88017 × DAS-59122-7 (*pat*, *CP4 epsps*, *cry1A.105*, *cry2Ab*, *cry3Bb1*, *cry34Ab1*, *cry35Ab1* and *cry1Fa2*) were granted for environmental impact in Canada, Japan and the USA in 2009, for food and/or feed in Mexico and the Philippines in 2010, and for food and feed use in Japan, Korea and Taiwan in 2009. Regulatory approval was granted for DAS-59122-7 × TC1507 × NK603 for the environment in Canada and Japan in 2006; for food and/or feed in Mexico in 2006; for food use in Japan in 2005, in Korea in 2006 and in the Philippines in 2007; feed use was approved in Japan in 2006 and the Philippines in 2007.

1.3.9 *CpTI* gene

Introduces resistance to insects

NOMENCLATURE: Approved name: *CpTI* gene.

Other names: *Cowpea trypsin inhibitor* gene.

Promoter(s): CaMV 35S.

SOURCE: The *cowpea trypsin inhibitor* gene was isolated from the cowpea (*Vigna unguiculata* L. Walp.), following observations that the bruchid beetle (*Callosobruchus maculatus* F.), feeding on stored cowpeas, failed to develop and often died.

PRODUCTION: The full-length cDNA clone encoding the trypsin inhibitor from cowpea was placed under the control of a CaMV 35S promoter and used in a construct for transfer into plants. The construct used the *Agrobacterium tumefaciens* Ti plasmid with kanamycin resistance as the selectable marker.

TARGET PESTS: Many insect pests, especially Lepidoptera and Coleoptera.

TARGET CROPS: Potatoes, oilseed rape, rice, strawberries, and other crops.

BIOLOGICAL ACTIVITY: Cowpea trypsin inhibitor is a protease that interferes with the digestion of insects. Trypsin is an important digestive enzyme, and its inhibition results in the failure of treated insects to incorporate consumed material into useable materials. **Mode of action:** The gene expressed in plants leads to the production of the trypsin inhibitor throughout the transgene. Insects that feed upon these plants have their gut trypsin inhibited, and this leads to a failure to grow and reproduce. Intoxicated insects cease to feed shortly after eating. **Efficacy:** Broad-spectrum insect control has been achieved with various transgenic crops. **Key reference(s):** V A Hilder, A M R Gatehouse, S E Sherman, R F Barker & D Boulter. 1987. A novel mechanism for insect resistance engineered into tobacco, *Nature*, **330**, 160–3.

COMMERCIALISATION: Formulation: The original work was undertaken at Durham University, UK, with support from Agricultural Genetics Company (now Pestax). Agreements for the development of CpTI have been signed with several companies, but no product has been commercialised. **Patents:** EP 272144; US 5218104 and US 4640836.

COMPATIBILITY: Transformed plants can be treated with conventional chemicals, and there is evidence that the use of *CpTI* genes as one component of a dual gene-insertion strategy gives more than additive insect control.

MAMMALIAN TOXICITY: There are no reports of allergic or other adverse toxicological properties associated with the use of *CpTI* genes by research workers, plant breeders or field workers.

ENVIRONMENTAL IMPACT AND NON-TARGET TOXICITY: Cowpea trypsin inhibitor occurs widely in Nature and it is not expected to have any adverse effects on non-target organisms that do not feed on GM crops or on the environment, when expressed in transgenic crops.

1.3.10 *GNA* gene

Introduces resistance to insects

NOMENCLATURE: Approved name: *GNA* gene.

Other names: *Snowdrop lectin* gene; *Galanthus nivalis agglutinin* gene.

Promoter(s): Cauliflower mosaic virus (CaMV) 35S promoter for overall expression in the plant and maize sucrose synthase 1 (Sh-1) promoter for expression in the phloem and, subsequently, for sucking insect control.

SOURCE: The snowdrop (*Galanthus nivalis* L.) lectin gene was originally identified and characterised at the Katholieke Universiteit, Leuven, Belgium. The lectin is found in most parts of the snowdrop, but is particularly abundant in the bulb. Lectins had been recognised as possessing insecticidal activity, but often this was associated with adverse mammalian toxicity. Collaboration with Agricultural Genetics Company (now Pestax) has led to the development of transgenic crop plants that contain the gene expressed constitutively or specifically in the phloem.

PRODUCTION: The gene was isolated as cDNA and cloned into *E. coli*. It was then cloned into an *Agrobacterium tumefaciens* plasmid containing the appropriate promoter. Plants were then transformed using classical transformation technology.

TARGET PESTS: Sucking insects, such as aphids and whitefly, and other phytophagous insects and nematodes.

TARGET CROPS: Potatoes, tobacco, tomatoes, maize (corn), rice, cereals and top fruit.

BIOLOGICAL ACTIVITY: Lectins exert their effects in insects and mammals by binding to glycoproteins in the gut of animals that consume them. The advantage of GNA is that it binds specifically to α-1,3-mannose residues, and these are not components of the mammalian gut. GNA is a tetrameric protein and does not bind to α-linked glucose.
Mode of action: The exact mode of action of GNA is not certain, but it is likely that the insecticidal effects are associated with the binding of the lectin to mannose residues in the

gut epithelial cells, leading to a disruption of epithelial cell function and death. **Efficacy:** Bioassays have shown that GNA is very active against members of the order Hemiptera, having an LC_{50} against the brown planthopper (*Nilaparvata lugens* (Stal.)) of 6 mM. **Key reference(s):** 1) AMR Gatehouse, RE Down, KS Powell, N Sauvion, Y Rahbe, CA Newell, A Merryweather, D Boulter & JA Gatehouse. 1996. Effects of GNA expressing transgenic potato plants on peach-potato aphid, *Myzus persicae*, *Entomol. Exp. Appl.*, **79**, 295-307. 2) WJ Peumans & EJM Van Damme. 1995. The role of lectins in plant defence, *Histochem. J.*, **27**, 253-71. 3) Y Shi, MB Wang, KS Powell, E Van Damme, VA Hilder, AMR Gatehouse, D Boulter & JA Gatehouse. 1994. Use of the rice sucrose synthase-1 promoter to direct phloem-specific expression of β-glucuronidase and snowdrop lectin genes in transgenic tobacco plants, *J. Exp. Bot.*, **45**, 623–31.

COMMERCIALISATION: Formulation: Pestax has entered a number of collaborative agreements for the development of GNA with several companies, but no product has been commercialised. **Patents:** US 5545820; Australia 668096.

COMPATIBILITY: Transgenic plants expressing the GNA gene can be treated with agrochemicals in the usual way. It is unlikely that insecticides will be needed. Transgenic plants containing GNA plus another gene coding for a different insect-specific toxin may be developed as a method for avoiding or delaying the onset of resistance.

MAMMALIAN TOXICITY: GNA is a mannose-specific binding lectin and, as such, is not considered to represent a toxicological hazard to mammals. There are no reports of allergic or other adverse toxic effects from research workers, breeders or field workers.

ENVIRONMENTAL IMPACT AND NON-TARGET TOXICITY: GNA occurs in Nature and its expression in transgenic crops is not expected to have any adverse effects on non-target organisms that do not feed on GM crops or on the environment.

1.3.11 *Bacillus thuringiensis AB88 vip3A(a)* gene

Introduces resistance to insects

NOMENCLATURE: Approved name: *Bacillus thuringiensis* AB88 *vip3A(a)* gene.

Other names: VipCot Cotton; COT102 × COT67B Cotton; Bt11 × MIR162 Corn.

Development code(s): SYN-IR1Ø2-7 (COT102); SYN-BTØ11-1, SYN-IR162-4 (BT11 × MIR162); BT11 × MIR162 × MIR604; SYN-IR162-4 (MIR162).

Promoter(s): COT102 – modified promoter, first exon and intron of *Arabidopsis thaliana* actin-2 gene. SYN-IR162-4, BT11 × MIR162 × MIR604 and MIR162 – ZmUbiInt (*Zea mays* polyubiquitin gene promoter and first intron.

Trait introduction: Trait introduction into COT102 and MIR162 was by *Agrobacterium tumefaciens*-mediated plant transformation. Trait introduction into BT11 × MIR162 and BT11 × MIR162 × MIR604 was by traditional plant breeding and selection.

SOURCE: The vegetative insecticidal protein VIP3A is the product of a gene originally derived from *Bacillus thuringiensis* Berliner strain AB88.

TARGET PESTS: COT102 is designed to control lepidopteran insect pests such as cotton bollworm (*Helicoverpa zea* (Boddie)), tobacco budworm (*Heliothis virescens* (Fabricius)), pink bollworm (*Pectinophora gossypiella* (Saunders)), fall armyworm (*Spodoptera frugiperda* (JE Smith)), beet armyworm (*Spodoptera exigua* (Hubner)), soybean looper (*Pseudoplusia includes* (Walker)), cabbage looper (*Trichoplusia ni* (Hubner)) and cotton leaf perforator (*Bucculatrix thurberiella* Busck.). In maize hybrids and stacked varieties, VIP3A(a) protein is very active against *H. zea*, *S. frugiperda*, black cutworm (*Agrotis ipsilon* Hufnagel) and western bean cutworm (*Richia albicosta* Smith) larvae. BT11 × MIR162 × MIR604 provides insect protection against European cornborer (*Ostrinia nubilalis* (Hubner)), southwestern cornborer (*Diatraea grandiosella* Dyar), southern cornstalk borer (*Diatraea crambidoides* (Grote)), corn earworm (*H. zea*), fall armyworm (*S. frugiperda*), armyworm (*Pseudaletia unipunctata* (Haworth) Franclemont), beet armyworm (*S. exigua*), black cutworm (*A. ipsilon*), western bean cutworm (*R. albicosta*), sugarcane borer (*Diatraea saccharalis* (Fabricius)), common stalk borer (*Papaipema nebris* (Guenée)), western corn rootworm (*Diabrotica virgifera virgifera* LeConte), northern corn rootworm (*Diabrotica barberi* Smith & Lawrence) and Mexican corn rootworm (*Diabrotica virgifera zeae* Krysan & Smith).

TARGET CROPS: COT102 – cotton. BT11 × MIR162, BT11 × MIR162 × MIR604 and MIR162 – maize (corn).

BIOLOGICAL ACTIVITY: Mode of action: The VIP3A proteins act by binding to specific sites located in the midgut epithelium of susceptible insect species of the order Lepidoptera. Following binding, cation-specific pores are formed that disrupt ion flow in the midgut, thereby causing paralysis and death. The insecticidal nature of the VIP3A protein is attributable to the presence of specific binding sites in these insects, which are different from those specific to Cry1Ab and Cry1Ac proteins.

COMMERCIALISATION: Trade names: 'Agrisure 3100' (*cry1A*(*b*) plus *vip3A*(*a*)) (BT11 × MIR162 × MIR604) (Syngenta Seeds), 'Agrisure 2100' (*cry1A*(*b*) plus *vip3A*(*a*) plus *pat*) (BT11 × MIR162) (Syngenta Seeds), 'VipCot' (vip3A(a)) (COT102) (Syngenta Seeds). **Patents:** COT102 – US Patent 7371940.

MAMMALIAN TOXICITY: Toxicological effects were determined from acute feeding studies. Leaf tissue and cottonseed from COT102 were not used in these studies to avoid possible confounding effects of the natural toxicants of cotton (such as gossypol). The organisms were fed VIP3A protein derived from either recombinant *E. coli*, or maize expressing the *vip3A*(*a*) gene. The VIP3A protein from both these sources was shown to be equivalent to that expressed by COT102 cotton, as demonstrated by similarity in molecular weight, immunoreactivity, results of mass spectral analysis of the peptides, and similarity in the rank-order of bioactivity against four species of VIP3A-sensitive lepidopteran larvae. The effect of VIP3A on mammals was investigated in an acute feeding study conducted with mice. VIP3A expressed in maize tissue was fed, by gavage, at 5000 mg/kg of body weight. VIP3A expressed in *E. coli* was also fed at 5050 mg/kg of body weight. No adverse effects were observed in the mice fed either diet during the 14 days of the acute feeding study. The digestibility of VIP3A in simulated gastric fluids was also investigated. The VIP3A protein expressed in *E. coli* was degraded, at time zero, to low molecular weight (<14000) peptides, and no peptides were detectable after 2 min. The VIP3A protein is, therefore, rapidly digested in a mammalian gut environment.

ENVIRONMENTAL IMPACT AND NON-TARGET TOXICITY: Most of the organisms in the acute feeding studies were fed, at the highest treatments, levels of VIP3A that were several times greater than the estimated environmental concentrations, based on the expression levels in COT102 tissues. No adverse effects were observed in any of the tested organisms at the highest treatment levels. For each of these organisms, the highest treatment levels represented the following exposure to VIP3A from COT102: bobwhite quail, 21-fold the level in leaves, and >1200-fold the level in cottonseed; honeybee larvae, >77-fold the level in pollen; adult honeybees, >8-fold the level in pollen; ladybird adults and lacewings, >133-fold the level in pollen; Collembola and earthworm, >2100- and 180-fold the estimated soil concentration, respectively. Environmental concentrations of VIP3A from COT102 were not estimated for either *Daphnia magna* or catfish. Aquatic organisms would most likely be exposed to VIP3A protein from COT102 pollen drifting from nearby fields and depositing on the surface of the water. This represents a very minimal amount of exposure since cotton flowers are mostly self-pollinating, and the pollen is heavy, thereby minimising dispersal by wind. No adverse effects were observed in either *Daphnia magna* or catfish when fed at the highest levels of the VIP3A protein.

DATE(S) OF REGISTRATION: COT102 achieved food use status in Australia and food and/or feed use in the USA in 2005; Mexico granted food and/or feed use in 2010. Registration of BT11 × MIR162 and BT11 × MIR162 × MIR604 corn was granted by US-EPA in 2009. MIR162 achieved regulatory approval for the environment in Brazil in 2009 and in Canada and the USA in 2010; food and/or feed approvals were granted in the USA in 2008, Brazil in 2009 and Mexico in 2010; food approvals were given in Australia and the USA in 2009 and in Canada, Japan and the Philippines in 2010; feed use approvals were granted in Canada and the Philippines in 2010.

2 OUTPUT TRAITS

2.1 *AAC synthase* gene

Reduces ethylene accumulation

NOMENCLATURE: Approved name: *1-amino-cyclopropane-1-carboxylic acid synthase* gene

Other names: *ACC synthase* gene.

Development code(s): Carnation – 66. Tomato – 1345-4, CGN-89322-3 (8338).

Promoter(s): ACC – CaMV 35S. ACCd – figwort mosaic virus (FMV) 35S5′ leader: petuniaHSP70 gene.

Trait introduction: Transgenic carnation line 66 was produced by *Agrobacterium*-mediated transformation of carnation plants (*Dianthus caryophyllus* L., cultivar Ashley) in which the transfer-DNA (T-DNA) region of the bacterial Ti plasmid was modified to contain a truncated copy of the ACC synthase-encoding gene and the ALS-encoding gene (*surB*) from a chlorsulfuron-tolerant isolate of tobacco. Line 66 was reproduced through vegetative reproduction, as were its progeny, as well as progeny derived through crosses with traditionally bred non-genetically modified carnation plants. The *surB* gene was included as a selectable marker. Tomato line 1345-4 was developed using recombinant DNA techniques to express the trait of delayed ripening of tomato fruit. The transgenic line contains a truncated version of the tomato *1-aminocyclopropane-1-carboxylic acid* (*ACC*) *synthase* gene, normally found in tomato. The *in situ* accumulation of ethylene in the transgenic tomatoes was only about 2% of the level found in the unmodified parental line, and the fruit does not fully ripen unless an external source of ethylene is applied. An antibiotic resistance marker gene (*neo*) encoding the enzyme neomycin phosphotransferase II (NPTII), which inactivates aminoglycoside antibiotics such as kanamycin and neomycin, was also introduced into the genome of tomato lines 1345-4 and 8338 as a selectable marker to identify transformed plants during tissue culture regeneration and multiplication. DR tomato line 8338 was developed by introducing into the genome of a processing tomato cultivar UC82B a gene (*ACCd*) derived from a non-pathogenic soil bacterium (*Pseudomonas chlororaphis* (Guignard & Sauvageau) Bergey *et al.*) that encodes the enzyme ACCd.

TARGET CROPS: Carnation, tomato

BIOLOGICAL ACTIVITY: Biology: Harvested carnation (*Dianthus caryophyllus* L.) flowers have a defined 'vase life', or time for which the flowers will stay fresh prior to beginning to shrink, dry out and die. This senescence process is triggered by production of a natural plant hormone, ethylene, by the flower. Flower death can also be triggered by exposure of flowers to external sources of ethylene, such as fruit and exhaust fumes. Typically, flower growers have controlled wilting by treating flowers with chemicals (e.g. silver thiosulfate) that either inhibit ethylene synthesis in the flower or prevent the flower from responding to ethylene. **Mode of action:** The enzyme 1-aminocyclopropane-1-carboxyllic acid (ACC) synthase, normally found in carnations, is responsible for the conversion of *S*-adenosylmethionine to ACC, which is the immediate precursor of ethylene. The transgenic carnation line 66 was developed using recombinant DNA techniques to display suppressed ACC synthase activity, and thus reduced ethylene synthesis and therefore longer vase life, by inserting an additional truncated copy of the *ACC synthase*-encoding gene. The presence of the truncated *ACC synthase* gene suppresses the normal expression of the native *ACC synthase* gene, and while not completely understood, the mechanism of 'downregulation' is probably linked to the coordinate suppression of transcription of both the endogenous gene and the introduced truncated *ACC synthase* gene. **Efficacy:** Carnation line 66 and derived lines have been tested in trials in the USA, the UK, the Netherlands and Australia. Results demonstrated that genetically modified carnations had a vase-life in water of 22 days, without requiring chemical treatment. In the same study, non-genetically modified carnations only had a vase-life in water of 10 days, and required chemical treatment to guarantee a vase-life in water of at least 8 days.

COMMERCIALISATION: Trade names: 'Carnation line 66' (*ACCase* plus *surB*) (Florigene), 'Tomato line 1345-4' (*ACCase* plus *neo*) (DNA Plant Technology), 'DR tomato line 8338' (*ACCase* plus *neo*) (Monsanto).

MAMMALIAN TOXICITY: There are no records of allergic or other adverse toxicological effects from researchers, breeders or growers of the transgenic varieties. The reduced synthesis of native ACC synthase was judged not to have any potential for additional human toxicity or allergenicity.

ENVIRONMENTAL IMPACT AND NON-TARGET TOXICITY: The risk of transferring genetic traits from transgenic carnation line 66 to species in unmanaged environments was not judged to be significant. The genetically modified carnation line 66 is intended for cultivation by cut flower growers, flower auctions, flower wholesalers, retailers and breeders. Plants are sold as flowers, cuttings or whole plants. The transgenic tomato line 1345-4 was field tested in the USA from 1992 to 1994. The agronomic characteristics of line 1345-4 were evaluated extensively in laboratory, glasshouse and field experiments. The line 1345-4 retained the agronomic characteristics of the parental tomato, and field reports on seed germination rates, yield characteristics, disease and pest susceptibilities, and fruit compositional analyses determined that line 1345-4 was comparable to the unmodified parental line, with the exception of reduced ACC synthase activity. Field trial reports demonstrated that transgenic tomato line 1345-4 did not exhibit weedy characteristics, and had no effect on non-target organisms or the general environment.

DATE(S) OF REGISTRATION: Carnation line 66 received environmental approval in Australia in 1995 and the EU in 1998; it also received marketing approval in the EU in 1998.

Environment regulatory approval for tomato line 1345-4 was granted in the USA in 1995; food and/or feed use was allowed in the USA in 1994 and in Mexico in 1998; food uses were approved in Canada in 1995. Food and/or feed regulatory approval for tomato line 8338 was allowed in the USA in 1994 and environmental approval was approved in 1995.

2.2 *barnase* and *barstar*

Introduces male sterility to plants

NOMENCLATURE: Approved name: *barnase* and *barstar*.

Development code(s): Canola – ACS-BNØØ4-7 × ACS-BNØØ1-4 (MS1, RF1 → PGS1); ACS-BNØØ4-7 × ACS-BNØØ2-5 (MS1, RF2 → PGS2); ACS-BNØØ5-8 × ACS-BNØØ3-6 (MS8 × RF3); PHY14, PHY35; PHY36. Chicory – RM3-3, RM3-4 and RM3-6. Maize (corn) – ACS-ZMØØ1-9 (MS3); ACS-ZMØØ5-4 (MS6).

Promoter(s): *barnase* pTa 29 pollen-specific promoter from *Nicotiana tabacum* L.; *barstar* anther-specific promoter.

SOURCE: The *barnase* gene is used for the introduction of male sterility and was isolated from *Bacillus amyloliquefaciens* (ex Fukumoto) Priest *et al*. The gene encodes for a ribonuclease enzyme (RNAse) expressed only in the tapetum cells of the pollen sac during anther development. The RNAse affects RNA production, disrupting normal cell functioning and arresting early anther development, thus leading to male sterility. The transgenic line RF1 (B93-101) was produced by genetically engineering plants to restore fertility in the hybrid line and to be tolerant to the herbicide glufosinate-ammonium (as a selectable marker). The *barstar* gene codes for a ribonuclease inhibitor (barstar enzyme) expressed only in the tapetum cells of the pollen sac during anther development. The ribonuclease inhibitor (barstar enzyme) specifically inhibits barnase RNAse. Together, the RNAse and the ribonuclease inhibitor form a very stable one-to-one complex, in which the RNAse is inactivated. As a result, when pollen from the restorer line is crossed to the male sterile line, the resultant progeny express the RNAse inhibitor in the tapetum cells of the anthers, allowing hybrid plants to develop normal anthers and restore fertility.

BIOLOGICAL ACTIVITY: The *barnase* gene encodes for a ribonuclease enzyme (RNAse) expressed only in the tapetum cells of the pollen sac during anther development. The RNAse affects RNA production, disrupting normal cell functioning and arresting early anther development, thus leading to male sterility. The *barstar* gene codes for a ribonuclease inhibitor (barstar enzyme) expressed only in the tapetum cells of the pollen sac during anther development. The ribonuclease inhibitor (barstar enzyme) specifically inhibits barnase RNAse. Together, the RNAse and the ribonuclease inhibitor form a very stable one-to-one complex, in which the RNAse is inactivated. As a result, when pollen from the restorer line is crossed to the male sterile line, the resultant progeny express the RNAse inhibitor in the tapetum cells of the anthers, allowing hybrid plants to develop normal anthers and restore fertility. **Biology:** The *barnase* and *barstar* genes are used by plant breeders to introduce male sterility to ensure outcrossing of new transgenic varieties.

2.3 *Bay TE* gene

Modifies seed fatty acid content

NOMENCLATURE: Approved name: *Bay TE.*

Other names: *thioesterase.*

Development code(s): 23-18-17 and 23-198.

Promoter(s): seed-specific promoter.

Trait introduction: The 23-198 and 23-18-17 lines were created by *Agrobacterium*-mediated transformation of line 212/86 in which the transfer-DNA (T-DNA) contained the gene encoding the enzyme 12:0 ACP thioesterase (bay TE) from the California bay tree (*Umbellularia californica* (Hook. & Arn.) Nutt.).

TARGET CROPS: Canola.

BIOLOGICAL ACTIVITY: The introduced gene encodes a thioesterase enzyme that is active in the fatty acid biosynthetic pathway of the developing seed, resulting in the accumulation of triacylglycerides containing esterified lauric acid (12:0) and, to a lesser extent, myristic acid (14:0). The processed oil derived from these novel varieties has a level of lauric acid similar to that of coconut ant palm kernel oil.

COMMERCIALISATION: Trade names: 'Laurical' (*Bay TE*) (Monsanto).

MAMMALIAN TOXICITY: There are no records of allergic or other adverse toxicological effects from researchers, breeders or growers of the transgenic variety.

ENVIRONMENTAL IMPACT AND NON-TARGET TOXICITY: The canola lines 23-198 and 23-18-17 were field tested in the USA (1991–92) and Canada (1993–95).

Agronomic and adaptation characteristics such as seed germination, seed yield, seedling growth, flowering and maturity dates were within the normal range of expression of characteristics in unmodified counterparts. The laurate canola lines were generally taller, later maturing and lower yielding, reflecting characteristics of the parent line 212/86. Disease incidence and insect susceptibilities were monitored during field trials and in glasshouses, and were shown to be the same as those of unmodified counterparts. The only significant consistent difference between high-laurate canola and the parent variety was the increase in laurate content from less than 0.1% to greater than 10%. The high-laurate canola lines are grown under contract to preserve their identity and prevent mixing with conventional canola. Overall, the field data reports demonstrated that canola lines 23-198 and 23-18-17 had no potential to pose a plant pest risk.

DATE(S) OF REGISTRATION: 'Laurical' canola was cleared for the environment in the USA in 1994 and Canada in 1996; it gained food and/or feed approval in the USA in 1994; and food and feed approval in Canada in 1996.

2.4 *cordapA* gene

Encodes for a lysine-insensitive dihydropicolinate synthase

2 OUTPUT TRAITS

NOMENCLATURE: Approved name: *cordapA* gene.

Development code(s): REN-ØØØ38-3 (LY038); REN-ØØØ38-3, MON-ØØ81Ø-6 (MON810 × LY038).

Promoter(s): *cordapA* – *globulin 1* (*Glb1*) gene from *Zea mays* rice actin gene intron (rAct1) and *Z. mays* chloroplast transit peptide sequence for DHDPS. *cryAb* – enhanced CaMV 35S, maize HSP70 intron.

Trait introduction: LY038 maize was developed using microprojectile bombardment with DNA-coated gold microprojectiles accelerated into callus derived from maize inbred line H99. The DNA used comprised a 5.9 Kb fragment of plasmid PV-ZMPQ76 containing the *cordapA* cassette and an *nptII* cassette as a selectable marker. The progeny were screened for plants that did not contain the *nptII* gene cassette, but contained the *cordapA* gene cassette. MON810 × LY038 maize is a stacked event resulting from the hybridisation of maize inbred MON 810 (MON-ØØ81Ø-6) with LY038 (REN-ØØØ38-3). This stacked maize hybrid is a product of traditional plant breeding, and therefore is not automatically subject to regulation in all jurisdictions as are transgenic plants resulting from recombinant DNA technologies.

SOURCE: The stacked hybrid MON810 × LY038 expresses the novel proteins expressed in the parental lines used to breed this line. The insecticidal delta-endotoxin Cry1Ab, which confers resistance to the European cornborer (*Ostrinia nubilalis* (Hübner)) and other lepidopterans, is produced by the *cry1Ab* gene from MON810. The *cordapA* gene is derived from *Corynebacterium glutamicum* (Kinoshita *et al.*).

TARGET CROPS: Maize (corn).

BIOLOGICAL ACTIVITY: LY038 maize has been genetically modified to express the *cordapA* gene from *C. glutamicum*. This gene encodes for a lysine-insensitive dihydropicolinate synthase (cDHDPS) enzyme, a regulatory enzyme in the lysine biosynthetic pathway. The activity of the native maize DHDPS is regulated by lysine feedback inhibition. Since the cDHDPS enzyme is less sensitive to lysine feedback inhibition, its expression in maize LY038 results in elevated levels of free lysine in the grain when compared with conventional maize. **Biology:** Maize–soybean meal-based diets formulated for poultry and swine are characteristically deficient in lysine and require the addition of supplemental lysine for optimal growth and production of these animals. The development of LY038 maize provides an alternative to direct supplementation of poultry and swine diets by increasing the amount of lysine in the maize portion of the feed.

COMMERCIALISATION: Trade names: 'LY038 Corn' (*cordapA*) (Monsanto), 'MON810 × LY038 Corn' (*cordapA* plus *cry1Ab*) (Monsanto).

MAMMALIAN TOXICITY: In *in vitro* digestion studies, greater than 96% of the cDHDPS protein was digested in simulated gastric fluid within 30 s, supporting the conclusion that cDHDPS has a low allergenic potential. An acute high-dose oral toxicity study in mice using *E. coli*-derived cDHDPS protein involving a single gavage dose of 800 mg of cDHDPS/kg body weight did not result in any mortalities or adverse clinical reactions in mice.

ENVIRONMENTAL IMPACT AND NON-TARGET TOXICITY: Numerous field trials were conducted in 2002 and 2003 in a variety of locations to evaluate LY038 maize. In each of these trials, LY038 was compared to a negative segregant (a sister line in which the *cordapA* transgene was not present), referred to as LY038(-), as well as several control and reference hybrids. Standard field trials included an evaluation of dormancy and germination, ecological evaluations (plant interactions with insect pests, disease and abiotic stresses), phenotypic evaluations and compositional changes. Data addressing the above categories were collected in order to assess possible effects from introduction of the *cordapA* gene and its associated regulatory sequences. No significant differences were found between the LY038 maize and the LY038(-) maize. It is expected that there will be no adverse environmental impact associated with the cultivation of LY038 maize.

DATE(S) OF REGISTRATION: Environment regulatory approvals for LY038 maize were granted in the USA and Canada in 2006 and in Japan in 2007; food and/or feed approval was obtained in the USA in 2005 and the Philippines in 2006; food uses were approved in Canada and Taiwan in 2006 and in Australia and Japan in 2007; and feed uses were approved in Canada in 2006 and in Japan and Mexico in 2007. Environment regulatory approval for MON810 × LY038 maize was granted in Japan in 2007; food and feed uses were approved in the Philippines in 2006 and in Japan in 2007.

2.5 *dfr*, *hfl* and *bp40* genes

Modifies flower colour

NOMENCLATURE: Approved name: *dfr*, *hfl* and *bp40* genes

Other names: *dihydroflavonol reductase* gene; *flavonoid 3p, 5p hydroxylase* gene; *flavonoid 3p, 5p hydroxylase* gene.

Development code(s): 4, 11, 15 and 16; 959A, 988A, 1226A, 1351A, 1363A and 1400A.

Promoter(s): *dfr* – *Agrobacterium tumefaciens*-constitutive mac-1 promoter; *hfl* – petal-specific CHS promoter derived form *Anthirrhinum majus*; *bp40* – petal-specific CHS promoter derived from *Petunia hybrida*; *surB* – CaMV 35S.

Trait introduction: Transgenic carnation lines 4, 11, 15 and 16 were developed using recombinant DNA techniques.

SOURCE: The variant form of ALS introduced into carnation lines 4, 11, 15 and 16 was isolated from a chlorsulfuron-tolerant tobacco (*Nicotiana tabacum* L.) and was introduced as a means of selecting for transformed plants during tissue culture regeneration as sulfonylurea herbicides are not, and will not be, used in the carnation industry.

TARGET CROPS: Carnation.

BIOLOGICAL ACTIVITY: Transgenic carnation lines 4, 11, 15 and 16 produce flowers with a unique violet/mauve colour by introducing two genes from petunia (*Petunia hybrida* L.) that function together in the biosynthesis of the anthocyanin pigment delphinidin. The transgenic lines were derived from the parent cultivar 'White Unesco', which is a white coloured carnation that was selected for a mutation in the dihydroflavonol reductase (DFR)-encoding gene that did not allow for expression of a functional enzyme, and thus did not produce the anthocyanin-type pigments that give rise to blue and red coloured flowers. The two genes introduced into the transgenic carnation lines included a functional dihydroflavonol reductase-encoding gene and a gene encoding the enzyme flavonoid 3′, 5′-hydroxylase (F3′, 5′H), a member of the NADPH-cytochrome P_{450} reductase family. Expression of the F3′, 5′H-encoding gene allows for the production of blue coloured delphinidin anthocyanin pigments, which are not normally found in carnation. Transgenic lines 959A, 988A, 1226A, 1351A, 1363A and 1400A were developed using recombinant DNA techniques to produce flowers with a unique deep purple colour, by introducing two genes that function together in the biosynthesis of delphinidin. The transgenic lines were derived from the parent cultivar 'White Unesco'. The two genes introduced into the transgenic carnation lines included a functional dihydroflavonol reductase-encoding gene isolated from petunia (*Petunia hybrida*) and a gene encoding F3′, 5′H isolated from *Viola*.

COMMERCIALISATION: Trade names: 'Carnation – 4, 11, 15, 16' (*surB* plus *dfr* plus *hfl*) and 'Carnation – 959A, 988A, 1226A, 1351A, 1363A and 1400A' (*surB* plus *dfr* plus *bp40*)' (Florigen).

MAMMALIAN TOXICITY: There are no records of allergic or other adverse toxicological effects from researchers, breeders or growers of the transgenic varieties.

ENVIRONMENTAL IMPACT AND NON-TARGET TOXICITY: The risk of transferring genetic traits from transgenic carnation lines 4, 11, 15 and 16 to species in unmanaged environments was not considered to be significant. The genetically modified carnation lines 4, 11, 15 and 16 are intended for cultivation by growers, flower auctions, flower wholesalers, retailers and breeders. Plants are sold as flowers, cuttings or whole plants.

DATE(S) OF REGISTRATION: Transgenic carnation lines 4, 11, 15 and 16 gained environment regulatory approval in Australia in 1995, and marketing approval in the EU in 1997. Environment regulatory approvals for transgenic carnation lines 959A, 988A, 1226A, 1351A, 1363A and 1400A were granted in the EU in 1998 and Colombia in 2000; marketing approval was given in the EU in 1998.

2.6 *fad2* gene

Modifies seed fatty acid content

NOMENCLATURE: Approved name: *fad2* gene.

Other names: *fatty acid desaturase* gene.

Development code(s): 45A37, 46A40, 46A12 and 46A16.

Trait introduction: Line 46A12 was derived from a single plant resulting from a cross between the high oleic acid parent, NS699, and NS1172, a broad-based spring canola population originating from European germplasm. Line 46A16 was derived from a cross between the high oleic acid parent, NS672, and NS1167, a broad-based population originating from Canadian germplasm. The processed oil derived from these novel varieties has levels of oleic acid similar to those of peanut and olive oils.

TARGET CROPS: Canola.

BIOLOGICAL ACTIVITY: The high oleic acid trait in lines 45A37, 46A40, 46A12 and 46A16 was selected following chemical mutagenesis by exposing seeds of canola varieties 'Regent', 'Topas' and 'Andor' to a solution of ethylnitrosourea (8 mM) in dimethylsulfoxide. Ethylnitrosurea is a commonly used chemical mutagen that affects DNA by chemically altering base pairs. It is believed that the induced mutation in lines 45A37 and 46A40 is analogous to that in *fad2* mutants of *Arabidopsis thaliana*. The *fad2* gene encodes a desaturase enzyme that catalyses the conversion of C18:1 to C18:2 and C18:3 fatty acids in plant cells. A mutation within the *fad2* gene that blocks expression of an active desaturase enzyme results in the accumulation of C18:1 oleic acid. The low linolenic acid trait was introduced by breeding with the registered low linolenic acid canola varieties 'Stellar' and 'Apollo'. The only significant difference between lines 45A37 and 46A40 is that the former was selected for early maturity, while 46A40 matures later in the growing season (medium maturity).

COMMERCIALISATION: Trade names: 'Stellar' and 'Apollo' (*fad2*) (Pioneer).

MAMMALIAN TOXICITY: There are no novel proteins produced and only the refined seed oil will be used as a food. Refined edible canola oil does not contain any detectable protein and consists of purified glycosides. Other than the traits of high oleic acid content and low linolenic acid content in the seed oil, the disease, pest and other agronomic characteristics of the 45A37 and 46A40 canola lines were comparable to other commercially available canola varieties.

ENVIRONMENTAL IMPACT AND NON-TARGET TOXICITY: 'Stellar' and 'Apollo' canola show great similarities to conventional varieties and are not expected to show any adverse effects on non-target organisms or the environment.

DATE(S) OF REGISTRATION: 45A37, 46A40, 46A12 and 46A16 canola received regulatory approval for food use in 1996.

2.7 *fan1* gene

Modifies seed fatty acid content

NOMENCLATURE: Approved name: *fan1* gene.

Development code(s): OT96-15.

Trait introduction: The OT96-15 soybean line was developed by traditional plant breeding methods using the variety 'Maple Glen' and PI361088B, a cultivar from Romania originating from Mica Ungana, as the source of the low linolenic acid trait.

TARGET CROPS: Soybean.

BIOLOGICAL ACTIVITY: The oil derived from OT96-15 has about one-half the content of linolenic acid compared with conventional soybean varieties, but has higher linolenic acid levels than either corn or olive oils.

COMMERCIALISATION: Trade names: 'Soybean OT96-15' (*fan1*) (Agriculture & Agri-Food Canada).

MAMMALIAN TOXICITY: There is no evidence that the OT96-15 line will have any adverse or allergenic effects.

ENVIRONMENTAL IMPACT AND NON-TARGET TOXICITY: The OT96-15 line shows great similarities to conventional soybeans and is not expected to show any adverse effects on non-target organisms or the environment.

DATE(S) OF REGISTRATION: Approved for food use in Canada in 2001.

2.8 *gmFAD2-1* gene

Modifies seed fatty acid content

NOMENCLATURE: Approved name: *gmFAD2-1* gene.

Other names: *delta(12)-fatty acid dehydrogenase* gene.

Development code(s): DD-Ø26ØØ5-3 (G94-1, G94-19 and G168).

Promoter(s): *gmFad2-1* – seed-specific promoter from the soybean beta-conglycinin gene; *dapA* – Kunitz trypsin inhibitor (Kti3) from soybean; *gus* – CaMV 35S; *bla* – bacterial promoter.

Trait introduction: The soybean sub-lines G94-1, G94-19 and G168 were produced via biolistic transformation of the parent soybean line (Asgrow A2396) with a mixture of two plasmids, pBS43 and pML102. These sub-lines were advanced using traditional breeding techniques to produce high oleic soybean varieties. The *bla* (*beta-lactamase*) gene confers resistance to the antibiotic ampicillin and permits the selection of transformed *E. coli* cells during laboratory recombinant DNA steps. The *bla* gene contains its own *E. coli* regulatory sequences and was therefore not expressed in the transformed plants.

TARGET CROPS: Soybean.

BIOLOGICAL ACTIVITY: Three lines of a new variety of soybean (G94-1, G94-19 and G168), high in the mono-unsaturated fatty acid oleic acid, were generated by the transfer of a second copy of a soybean fatty acid desaturase gene (*GmFad2-1*) to a high-yielding commercial variety of soybean (line A2396). The fatty acid desaturase is responsible for the synthesis of linoleic acid, which is the major polyunsaturated fatty acid present in soybean oil. The presence of a second copy of the fatty acid desaturase gene causes a phenomenon known as 'gene silencing', which results in both copies of the fatty acid desaturase gene being 'switched off', thus preventing linoleic acid from being synthesised and leading to the accumulation of oleic acid in the developing soybean seed. The *dapA* gene is from *Corynebacterium glutamicum* (Kinoshita *et al.*), which encodes the enzyme dihydrodipicolinic acid synthase (DHDPS). *Corynebacterium* DHDPS catalyses a step in the biosynthesis of the amino acid lysine, but is insensitive to feedback inhibition by lysine. Expression of this enzyme in soybean seeds results in accumulation of the essential amino acid lysine.

COMMERCIALISATION: Trade names: 'High Oleic Acid Soybean' (*gmFad2-1* plus *dapA*) (Pioneer) and (DuPont).

MAMMALIAN TOXICITY: Feeding studies carried out with pigs and chickens demonstrated that processed soybean meal derived from high oleic soybeans was nutritionally equivalent to processed soybean meal derived from the conventional soybeans.

It was determined that high oleic soybeans may be used as any other non-modified soybean meal and would be treated as any other commodity soybean meal. Allergenicity studies demonstrated that there were no significant quantitative or qualitative differences between the high oleic soybean and non-modified elite soybean. The allergenicity potential was the same for transgenic soybean sub-lines and elite soybean with regard to their allergen content.

ENVIRONMENTAL IMPACT AND NON-TARGET TOXICITY: High oleic soybean lines derived from sub-lines G94-1, G94-19 and G168 were field tested in the USA (1995–96), Canada, Puerto Rico and Chile. Field test data concerning yields and visual observations of agronomic properties, including susceptibility to diseases and insects, indicated that high oleic soybeans were within the range normally displayed by non-modified varieties. Field observations revealed no negative effects on non-target organisms, suggesting that the relatively higher levels of the oleic acid in the tissues of these sub-lines are not toxic to organisms. Indirect metabolic alterations caused by the genetic modification were assessed and determined to have no impact on non-target organisms. Furthermore, the lack of known toxicity of oleic acid suggests no potential for deleterious effects on beneficial organisms. It was determined that genetically modified soybean lines G94-1, G94-19 and G168 did not have a significant adverse impact on organisms beneficial to plants or agriculture, or on non-target organisms, and were not expected to impact on threatened or endangered species.

DATE(S) OF REGISTRATION: Soybean sub-lines G94-1, G94-19 and G168 gained environmental regulatory approval in the USA in 1997, Japan in 1999 and Canada in 2000; food and/or feed clearance was obtained in the USA in 1997; food use approvals were granted in Australia and Canada in 2000 and in Japan in 2001; feed uses were approved in Canada and Japan in 2000.

2.9 *gm-fad2-1* gene

Modifies seed fatty acid content

NOMENCLATURE: Approved name: *gm-fad2-1* gene.

Other names: *delta(12)-fatty acid dehydrogenase* gene.

Development code(s): DP-3Ø5423-1 (DP-305423).

Promoter(s): *gm-fad2-1* – Kti3 (*Glycine max* Kunitz trypsin-inhibitor gene promoter); *gm-hra* – *S*-adenosyl-L-methionine synthetase (SAMS).

Trait introduction: Microprojectile bombardment was used to co-transform secondary plant cell embryos with two purified linear DNA fragments: a 2924 base pair fragment (PHP19340A fragment) containing the *gm-fad2-1* cassette and the 4512 base pair fragment (PHP17752A fragment) containing the *gm-hra* cassette.

TARGET CROPS: Soybean.

BIOLOGICAL ACTIVITY: The intended effect of the modification in 305423 soybean is to produce soybean seeds with increased levels of mono-unsaturated fatty acid (oleic) and decreased levels of polyunsaturated fatty acids (linoleic and linolenic). The *gm-hra* gene was used as a selectable marker; it confers tolerance to ALS-inhibiting herbicides.

COMMERCIALISATION: Trade names: 'Soybean DP-305423' (*gmfad2-1* plus *gm-hra*) (Pioneer).

MAMMALIAN TOXICITY: A study was conducted to assess whether the transformation process may have increased the overall allergenicity of 305423 soybean compared with conventional soybean. Using sera from clinically reactive soy-allergic patients, IgE immunoblot and ELISA studies were conducted using protein extracts from 305423 soybean and conventional soybean. The SDS-PAGE Coomassie blue-stained protein profiles for 305423 and control soybean extracts appeared to be the same; they are similar in their IgE-binding profile, and showed the same IgE-binding capacity for 305423 and control soy extracts. It was concluded that the levels of endogenous allergens in, and the allergic potential of, 305423 soybean are comparable to those in non-transgenic control soybean. The only newly expressed protein in 305423 soybean is the GM-HRA protein. Based on amino acid sequence similarity searches against the National Center for Biotechnology Information (NCBI) Protein dataset, there were no significant similarities with known protein toxins. An acute oral toxicity study was conducted in mice in which a single dose of 582 mg per kilogram of body weight (mg/kg bw) of GM-HRA protein was administered by oral gavage to five male and five female mice. No clinical symptoms of toxicity, body weight loss, gross organ lesions or mortality were observed. The result of this study shows that the GM-HRA protein does not cause acute toxicity.

ENVIRONMENTAL IMPACT AND NON-TARGET TOXICITY: A poultry feeding study over a period of 42 days was carried out with diets containing 305423 soybean, non-GM control soybean with comparable genetic background and three commercial non-GM soybeans. The mortality, body weight gain, feed efficiency, organ yield, carcass yield, breast, thigh, wing and leg yield and abdominal fat of the chickens fed with 305423 soybean were compared with chickens fed non-GM control soybean diets. The results from this 42-day poultry study confirm the safety of 305423 soybean and that 305423 soybean is nutritionally equivalent to non-GM control soybean with a comparable genetic background and to commercial soybeans.

DATE(S) OF REGISTRATION: Environment regulatory approval was granted in Canada in 2009 and the USA in 2010; food and/or feed approval was granted in Mexico in 2008 and in the USA in 2009; food uses were approved in Canada in 2009 and in Australia in 2010; feed uses were approved in Canada in 2009.

2.10 *Nicotinate-nucleotide pyrophosphorylase* gene

Reduces nicotine content in tobacco

NOMENCLATURE: Approved name: *Nicotinate-nucleotide pyrophosphorylase* gene

Development code(s): Vector 21-41.

Promoter(s): NtQPT1.

Trait introduction: Tobacco line Vector 21-41 was produced by *Agrobacterium*-mediated transformation of leaf discs from 'Burley 21 LA' using the binary vector pYTY32 which carried the NtQTP1 promoter and the antisense configuration of NtQTP1 cDNA.

TARGET CROPS: Tobacco.

BIOLOGICAL ACTIVITY: Tobacco line Vector 21-41 was developed to reduce the presence of nicotine in the leaves of commercial tobacco (*Nicotiania tabacum* L.). Vector 21-41 delivers nicotine levels around 20 times less than conventional tobacco. Nicotine is commonly regarded as the addictive substance in tobacco and is a precursor of tobacco-specific nitrosamines, which are mutagenic and carcinogenic substances normally found in tobacco. It is thought that delivering nicotine at levels well below those determined to be addictive can reduce dependency on tobacco products. **Mode of action:** The synthesis of nicotine is isolated in the roots of tobacco and subsequently transported via the phloem to the leaves. Nicotine is the product of two enzymatic pathways: the methylpyrroline pathway, which produces the nicotine precursor *N*-methylpyrrolinium cation; and the pyridine nucleotide cycle, which produces the precursor nicotinic acid. In the methylpyrroline pathway, L-ornithine is converted via ornithine decarboxylase to putrescine, which in turn is acted on by putrescine *N*-methyltransferase (PMTase) to produce 4-methyl putrescine and subsequently *N*-methylpyrrolinium cation. PMTase is the limiting enzyme in this pathway. The pyridine nucleotide cycle begins with 3-phosphoglyceraldehyde, which condenses with aspartic acid to form quinolinic acid. Quinolinic acid is subject to the action of quinolinic acid phosphoribosyltransferase (QPTase) and through a series of steps becomes nicotinic acid. In this pathway, QPTase is the limiting enzyme. In Vector 21-41, the pyridine nucleotide cycle was downregulated through the insertion of the antisense configuration of the gene coding for QPTase (*NtQTP1*). The reduction in expression of QPTase resulted in a dramatic decrease in the production of nicotinic acid and thus nicotine.

COMMERCIALISATION: Trade names: 'Vector 21-41 Tobacco' (Vector Tobacco) (not commercialised).

MAMMALIAN TOXICITY: Vector 21-41 was no more toxic to mammals than non-transformed tobacco.

ENVIRONMENTAL IMPACT AND NON-TARGET TOXICITY: The presence of nicotine in tobacco is thought to have evolved as a defence against insect predation. It is known that insects have developed resistance to nicotine, resulting in the need for multiple applications of pesticides on conventional tobacco crops. Vector 21-41 was found to have similar yields to other low-alkaloid tobacco cultivars when a standard package of pesticides was applied. With no insect management, both Vector 21-41 line and conventional tobacco had no commercial yield when faced with heavy insect pressure. Vector 21-41 did not result in increased pesticide application.

DATE(S) OF REGISTRATION: Environment regulatory approval was granted in the USA in 2002.

2.11 *polygalacturonase* gene

Suppresses polygalacturonase enzyme activity

NOMENCLATURE: Approved name: *polygalacturonase* gene.

Other names: *PG*.

Development code(s): B, Da, F; CGN-89564-2 ('FlavrSavr').

Promoter(s): CaMV 35S.

Trait introduction: The transgenic tomato lines B, Da and F were produced via *Agrobacterium*-mediated transformation of the inbred tomato line TGT7 (T7) (*Lycopersicon* *esculentum* L.). This transformation system was based on the disarmed binary vector Bin19, which was engineered to contain DNA sequences encoding a partial copy of the *PG* gene. During transformation, the T-DNA portion of the plasmid was transferred into the plant cells and stably integrated into the plant's genome. The PG-encoding gene was isolated from the tomato cultivar 'Ailsa Craig' and was inserted into the Ti binary vector in either the sense or antisense orientation. Tomato line F was used to develop the tomato hybrids 1401F, H282F, 11013F and 7913F. 'FlavrSavr' tomatoes were developed by insertion of an additional copy of the *polygalacturonase* (*PG*)-encoding gene in the 'antisense' orientation, resulting in reduced translation of the endogenous PG messenger RNA (mRNA).

TARGET CROPS: Tomato.

BIOLOGICAL ACTIVITY: The tomato lines B, Da and F contain a partial *polygalacturonase* (*PG*) gene that encodes for the PG protein, a pectin-degrading enzyme derived from tomato. The lines differ slightly in that Da and F contain the partial *PG* gene in the sense orientation while line B contains a partial antisense *PG* gene, essentially a reverse copy. The presence of the partial *PG* gene, in either sense or antisense orientation, suppresses the expression of endogenous PG enzyme at the onset of fruit ripening. In the case of line B, the mechanism of action is probably linked to the hybridisation of antisense and sense mRNA transcripts, resulting in a decreased amount of free positive sense mRNA available for protein translation. For lines Da and F, which contain the truncated *PG* gene in the sense orientation, reduced *PG* expression may be due to coordinate suppression of transcription of both the endogenous gene and the introduced truncated gene. Reduced *PG* expression decreases the breakdown of pectin and leads to fruit with slowed cell wall breakdown, better viscosity characteristics and delayed softening. Tomato lines B, Da and F have improved harvest and processing properties that allow the transgenic tomatoes to remain longer on the vine to develop their natural flavour, maintain their firmness for shipping and produce a thicker consistency in processing. 'FlavrSavr' tomatoes were developed using recombinant DNA techniques to express the trait of delayed softening of fruit.

COMMERCIALISATION: Trade names: 'Tomato lines B, Da and F' (*PG*) (Zeneca (now Syngenta)) (withdrawn), 'FlavrSavr' (Calgene) (withdrawn).

MAMMALIAN TOXICITY: The reduced synthesis of native PG was not judged to have any potential for human toxicity or allergenicity. Polygalacturonase is a natural component of all food plants. The partial sense *PG* gene present in transformation event F was identical to the corresponding endogenous tomato gene and conferred no additional risks. The antisense *PG* gene did not encode for any new protein products, and therefore should not have any toxic properties. The reduced synthesis of native PG in 'FlavrSavr' tomatoes was not judged to have any potential for human toxicity or allergenicity. Additional data from three 28-day rat feeding trials demonstrated no biologically significant changes in body weight, organ weight, food consumption, haematologic parameters, and clinical chemistry findings. There was disparity among the three studies regarding the incidence of rats with gastric erosions. Although no definitive conclusions could be drawn regarding the etiologies of these erosions, it was determined that they were also present in rats fed non-transgenic tomatoes and that, where they occurred, they were no more severe in rats fed 'FlavrSavr' tomatoes than in rats fed non-transgenic tomatoes.

ENVIRONMENTAL IMPACT AND NON-TARGET TOXICITY: The reduced levels of PG enzyme in genetically modified lines B, Da, and F should not have any toxic properties. Tomato flowers are unattractive to insect pollinators due to the limited availability of pollen. Bumblebee (*Bombus terrestris* (L.)) activity was found to be no different on modified and non-modified plants. Field testing determined that there were no significant differences in the susceptibility of modified and non-modified crops to pests and diseases. It was concluded that the genes inserted into transgenic tomato lines B, Da and F would not result in any deleterious effects or significant impacts on non-target organisms, including those that are recognised as beneficial to agriculture and those that are recognised as threatened or endangered in the USA. 'FlavrSavr' tomatoes were field tested in the USA from 1988 to 1992. Field trials demonstrated that the variation in agronomic characteristics among the tomato lines did not differ significantly from the natural variation found in commercial cultivars of tomato. These reports determined that 'FlavrSavr' tomatoes had no effect on non-target organisms or the general environment.

DATE(S) OF REGISTRATION: Environment regulatory approval was granted for lines B, Da and F in the USA in 1995; food and/or feed uses were approved in the USA in 1994 and in Mexico in 1996; and food uses were approved in Canada in 1996. Environment regulatory approvals were granted for 'FlavrSavr' tomatoes in the USA in 1992, Mexico in 1995 and Japan in 1996; food and/or feed approvals were granted in the USA in 1994; food uses were approved in Canada and Mexico in 1995 and in Japan in 1997; feed use approval was granted in Mexico in 1995.

2.12 *sam-k* gene

Delays fruit ripening

NOMENCLATURE: Approved name: *sam-k* gene.

Other names: *S-adenosylmethionine hydrolase* gene.

Development code(s): Cantaloupe – A and B; tomato – 35 1 N.

Trait introduction: The cantaloupe lines A and B were created by *Agrobacterium*-mediated transformation in which the transfer-DNA (T-DNA) contained the S-adenosylmethionine hydrolase encoding the *sam-k* gene from *E. coli* bacteriophage T3. The constitutive expression of the *sam-k* gene was controlled by inclusion of regulatory DNA sequences from *Agrobacterium tumefaciens*.

TARGET CROPS: Cantaloupe and tomato.

BIOLOGICAL ACTIVITY: The A and B lines of cantaloupe (*Cucumis melo* L.) were developed through a specific genetic modification to express a reduced accumulation of S-adenosylmethionine (SAM) and consequently the trait of delayed ripening. The conversion of SAM to 1-aminocyclopropane-1-carboxylic acid (ACC) is the first step in ethylene biosynthesis, and the lack of sufficient pools of SAM results in significantly reduced synthesis of ethylene, which is known to play a key role in fruit ripening. 'Delayed Ripening tomato line 35 1 N' was developed by using standard *Agrobacterium* binary vectors for introducing into the genome of 'Large Red Cherry' an *S-adenosylmethionine hydrolase* (*SAMase*)-encoding gene derived from *E. coli* bacteriophage T3. This results in transformed cherry tomato plants that exhibit significantly reduced levels of S-adenosylmethionine (SAM).

COMMERCIALISATION: Trade names: 'Cantaloupe lines A and B' (*sam-k*) (Agritope) (withdrawn), 'Delayed Ripening tomato line 35 1 N' (*sam-k*) (Agritope) (withdrawn).

MAMMALIAN TOXICITY: There are no records of allergic or other adverse toxicological effects from researchers, breeders or growers of the transgenic varieties.

ENVIRONMENTAL IMPACT AND NON-TARGET TOXICITY: These transgenic lines show great similarities to conventional crops and are not expected to show any adverse effects on non-target organisms or the environment.

DATE(S) OF REGISTRATION: 'Cantaloupe lines A and B' were approved for food uses in the USA in 1999. Environment and food and/or feed regulatory approval was granted in the USA for 'Delayed Ripening tomato line 35 1 N' in 1996.

2.13 Thermostable alpha-amylase

Provides heat-stable alpha-amylase

NOMENCLATURE: Approved name: Thermostable alpha-amylase enzyme.

Other names: AMY797E.

Development code(s): Maize line 3272.

Promoter(s): *amy797E* – GZein promoter – from maize 27-kDa *zein* gene PEPC9 – intron 9 from maize phospho-enol-pyruvate carboxylase gene.

Trait introduction: Maize line 3272 was produced by *Agrobacterium*-mediated transformation of immature embryos derived from the proprietary line A188 of *Zea mays* (maize) using the transformation vector pNOV7013. The T-DNA segment of this plasmid contained, between the right and left borders, the *amy797E* and *pmi* genes and their regulatory elements. The chimeric *amy797E* gene was generated to combine the best features of three thermostable amylase enzymes isolated from *Thermococcus* spp. bacteria, designated as BD5031, BD5064 and BD5063. Fragments from the three parental genes were combined (in the same relative position) based on the natural sequence homology to create a library of recombinant alpha-amylase genes. The chimeric AMY797E alpha-amylase enzyme (alternatively known as '797GL3') was identified by screening these recombined enzymes and is composed of four fragments from BD5031, two fragments from BD5064 and three fragments from BD5063. The *pmi* gene represents the *manA* gene from *E. coli* and encodes the enzyme phosphomannose isomerase (PMI). It was used as a selectable marker gene during the transformation process.

BIOLOGICAL ACTIVITY: Maize line 3272 has been genetically modified to express a thermostable alpha-amylase enzyme (AMY797E) for use in dry-grind fuel ethanol production in the USA. Amylase is used to hydrolyse starch into smaller sugar subunits, which is the first step in producing ethanol from plants. Amylases used for ethanol production need to be able to work at high temperatures and low calcium concentrations. Plants such as maize naturally contain amylases; however, these are destroyed when maize is subjected to the high processing temperatures necessary for ethanol production, making it necessary to add microbially produced amylase preparations. The use of Event 3272, expressing a highly thermostable amylase, bypasses this step.

COMMERCIALISATION: Trade names: 'Event 3272 Corn' (*amy797E* plus *pmi*) (Syngenta).

MAMMALIAN TOXICITY: The AMY797E protein is not homologous to known toxins. A single-dose oral toxicity study showed that the AMY797E protein is not toxic to the mouse at a dose of 1511 mg/kg body weight. This dose represents about 2.1 times the worst-case daily dietary dose for rodents eating a diet comprising 100% kernels of maize Event 3272 in the field. Alpha-amylases of varying degrees of amino acid homology with AMY797E protein occur widely in nature among prokaryotes and eukaryotes. Alpha-amylase enzymes are present in plants, including maize. Alpha-amylase enzymes are also found in human saliva. There is an enormous diversity of alpha-amylases in soil microorganisms, including many heat-stable alpha-amylases. Therefore, it is likely that mammals, birds, insects and microorganisms exposed to AMY797E protein expressed in Event 3272 have had prior exposure to alpha-amylases. No harmful effects of such exposure are known.

ENVIRONMENTAL IMPACT AND NON-TARGET TOXICITY: Event 3272 maize hybrids were tested at eight locations in 2003 and 17 locations in 2004 in the US corn belt. A total of 26 agronomic traits were evaluated. These agronomic traits covered a broad range of characteristics that encompass the entire life cycle of the maize plant and included data assessing seedling emergence, vegetative vigour, basic morphology, growth habit, time to reproduction, susceptibility to insect and disease attack, and yield characteristics. Event 3272 is not expected to exhibit modified response to diseases and pest insects, compared with unmodified corn counterparts, as the expression of the AMY797E alpha-amylase and PMI phosphomannose isomerase is unrelated to plant pest potential. For the majority of agronomic traits, no statistically significant differences between Event 3272 hybrids and their non-transformed isogenic counterparts were observed. A 42-day feeding study in broiler chickens showed no evidence of biologically significant differences in growth or feed conversion in chickens fed Event 3272 grain compared with near-isogenic control or commercially available maize lines.

DATE(S) OF REGISTRATION: Environmental regulatory approval for SYN-E3272-5 maize was allowed in Canada in 2008; food and/or feed uses were allowed in the USA in 2007 and in Australia, Mexico and the Philippines in 2008; food uses were approved in Canada in 2008 and in Russia in 2010; feed uses were allowed in Canada in 2008.

Glossary: Latin–English

Names include species referred to in the Main Entries and other agriculturally important organisms, some of which are not specifically mentioned within a main entry. For each name that is identified in the first column at the Genus level (i.e. names in italics), the third column gives: for fungi, bacteria, insects and vertebrates, the Order then Family; for plants, the Family. The first column also includes some Families and Orders, with a corresponding higher level indicated in the third column. Wherever possible, the authorities for these Latin names are also given.

Latin	English	Order and/or Family
Abutilon theophrasti Medic.	Velvetleaf	Malvaceae
Acantholyda erythrocephala (L.)	Pine false webworm	Hymenoptera: Pamphiliidae
Acari	Mites	
Acaridae	Acarid mites	Acari
Acarina (see Acari)		
Acarus siro L.	Storage mite	Acarina: Acaridae
Aceria guerreronis Keifer	Coconut mite	Acari: Eriophyidae
Acrididae	Grasshoppers and locusts	Saltatoria
Actinomycetales	Filamentous bacteria	
Aculops spp.	Mites	Acari: Eriophyidae
Aculus spp.	Rust mites	Acari: Eriophyidae
Aculus schlechtendali (Nalepa)	Rust mite, apple	Acari: Eriophyidae
Acyrthosiphon kondoi Shinji	Alfalfa aphid	Hemiptera: Aphididae
Acyrthosiphon pisum (Harris)	Pea aphid	Hemiptera: Aphididae
Adoryphorus couloni (Burmeister)	Redheaded cockchafer	Coleoptera: Scarabaeidae
Adoxophyes spp.	Tortrix moths; leaf rollers	Lepidoptera: Tortricidae
Adoxophyes orana Fischer von Roeslerstamm	Summer fruit tortrix moth	Lepidoptera: Tortricidae
Aedes aegypti (L.)	Yellow fever mosquito	Diptera: Culicidae
Aeschynomene spp.	Joint vetches	Fabaceae
Aeschynomene virginica L.	Northern joint vetch	Fabaceae
Agaricales	Mushrooms, etc.	
Agriotes spp.	Wireworms	Coleoptera: Elateridae
Agrobacterium radiobacter (Beijerink and van Delden)	Beneficial bacterium	Eubacteriales: Rhizobiaceae
Agrobacterium tumefasciens Conn.	Crown gall	Eubacteriales: Rhizobiaceae
Agromyza spp.	Leaf miners	Diptera: Agromyzidae
Agropyron repens Beauv. (see *Elytrigia repens*)		

References

Latin	**English**	**Order and/or Family**
Agrostis gigantea Roth	Black bent	Poaceae
Agrostis stolonifera L.	Creeping bent	Poaceae
Agrotis spp.	Cutworms	Lepidoptera: Noctuidae
Agrotis ipsilon (Hufnagel)	Black cutworm	Lepidoptera: Noctuidae
Agrotis segetum (Denis & Schiffermlller)	Turnip moth	Lepidoptera: Noctuidae
Alabama argillacea (Hübner)	Cotton leaf worm	Lepidoptera: Noctuidae
Albugo candida Kuntze	White blister	Oomycetes: Peronosporales
Aleurothrixus floccosus (Mask.)	Whitefly	Hemiptera: Aleyrodidae
Aleyrodidae	Whiteflies	Hemiptera
Alopecurus myosuroides Huds.	Black-grass	Poaceae
Alphitobius spp.	Mealworms	Coleoptera: Tenebrionidae
Alternanthera philoxeroides (Mart.) (Griseb.)	Alligatorweed	Amaranthaceae
Alternaria spp.	Leaf spots, various	Deuteromycetes: Moniliales
Alternaria alternata	Leaf spot	Deuteromycetes: Moniliales
Alternaria brassicae Sacc.	Dark leaf spot, brassicas	Deuteromycetes: Moniliales
Alternaria brassicicola Wiltsh.	Dark leaf spot, brassicas	Deuteromycetes: Moniliales
Alternaria dauci Groves & Skolko	Carrot leaf blight	Deuteromycetes: Moniliales
Amaranthus spp.	Amaranths	Amaranthaceae
Amaranthus retroflexus L.	Redroot pigweed; common amaranth	Amaranthaceae
Amblyseius barkeri (Hughes)	Predatory mite	Mesostigmata: Phytoseiidae
Amblyseius californicus (McGregor)	Predatory mite	Mesostigmata: Phytoseiidae
Amblyseius cucumeris (Oudemans)	Predatory mite	Mesostigmata: Phytoseiidae
Amblyseius degenerans (Berlese)	Fruit tree red spider mite predator	Mesostigmata: Phytoseiidae
Amblyseius fallacis (Garman)	Spider mite predator	Mesostigmata: Phytoseiidae
Ambrosia artemisifolia L.	Ragweed, common	Asteraceae
Ampelomyces quisqualis	Hyperparasite of fungi	Deuteromycetes: Sphaeropsidales
Amylois transitella (Walker)	Navel orangeworm	Lepidoptera: Pyralidae
Anagrapha falcifera (Kirby)	Celery looper; alfalfa looper	Lepidoptera: Noctuidae
Anagrus atomus (L.)	Leafhopper egg parasitoid	Hymenoptera: Mymaridae
Anarsia lineatella Zeller	Peach tree borer	Lepidoptera: Gelechiidae
Anopheles spp.	Mosquitos	Diptera: Culicideae
Anthomyiidae, *Delia* spp. (= some *Hylemya* spp.) and others	Root flies or maggots	Diptera: Anthomyiidae
Anthonomus grandis Boheman	Cotton boll weevil	Coleoptera: Curculionidae
Anticarsia gemmatalis Hübner	Soybean looper; velvet bean caterpillar	Lepidoptera: Noctuidae
Aonidiella aurantii (Maskell)	Californian red scale	Hemiptera: Diaspididae
Aonidiella orientalis (Newstead)	Oriental red scale	Hemiptera: Diaspididae
Apera spica-venti Beauv.	Loose silky-bent	Poaceae

Latin	English	Order and/or Family
Aphanomyces spp.	Foot rot; root rot, various hosts	Oomycetes: Saprolegniales
Aphanomyces cochlioides Drechs.	Blackleg, beet crops	Oomycetes: Saprolegniales
Aphelinus spp.	Aphid parasitoid wasps	Hymenoptera: Aphelinidae
Aphididae	Aphids	Hemiptera
Aphidius spp.	Aphid parasitoid wasps	Hymenoptera: Aphelinidae
Aphidoletes aphidimyza Rondani	Aphid gall midge	Diptera: Cecidomyiidae
Aphis citricida (Kirkaldy)	Black citrus aphid	Hemiptera: Aphididae
Aphis fabae Scopoli	Black bean aphid	Hemiptera: Aphididae
Aphis gossypii Glover	Melon aphid; cotton aphid	Hemiptera: Aphididae
Aphyllophorales		Basidiomycotina
Aphytis lignanensis Compére	Red scale parasite	Hymenoptera: Aphelinidae
Aphytis melinus DeBach	Golden chalcid	Hymenoptera: Aphelinidae
Archips podanus (Scopoli)	Leaf roller	Lepidoptera: Tortricidae
Armillaria mellea Kumm.	Honey fungus	Basidiomycetes: Agaricales
Arrhenatherum elatius Beauv.	False oat-grass	Poaceae
Arrhenatherum elatius var. *bulbosum* Spenn.	Onion couch	Poaceae
Artemisia vulgaris L.	Mugwort; wormwood	Asteraceae
Ascochyta spp.	Leaf spots, various hosts	Deuteromycetes: Sphaeropsidales
Ascochyta chrysanthemi Stevens (see *Didymella ligulicola*)		
Ascochyta fabae Speg.	Leaf spot, bean	Deuteromycetes: Sphaeropsidales
Ascochyta pinodes Jones	Leaf and pod spot, pea	Deuteromycetes: Sphaeropsidales
Ascochyta pisi Lib.	Leaf and pod spot, pea	Deuteromycetes: Sphaeropsidales
Ascomycotina	Fungi, sexually produced spores in sacs	
Aspergillus spp.	Storage fungi	Deuteromycetes: Moniliales
Aspidiotus nerii (Bouché)	Red scale	Hemiptera: Diaspididae
Athous spp.	Garden wireworms	Coleoptera: Elateridae
Atomaria linearis Stephens	Pygmy mangold beetle	Coleoptera: Cryptophagidae
Atriplex patula L.	Common orache	Chenopodiaceae
Aulacorthum solani (Kaltenbach)	Glasshouse potato aphid	Hemiptera: Aphididae
Autographa californica (Speyer)	Alfalfa looper	Lepidoptera: Noctuidae
Avena spp.	Oats (wild and cultivated)	Poaceae
Avena barbata Brot.	Bearded oat	Poaceae
Avena fatua L.	Wild oat	Poaceae
Avena sterilis L.	Oat, sterile	Poaceae
Avena sterilis L. ssp. *Ludoviciana* (= *A. ludoviciana* Durieu)	Wild oat, winter	Poaceae

Latin	English	Order and/or Family
Azadirachta indica A. Juss.	Neem tree	Meliaceae
Bacillus sphaericus Meyer & Neide		Schizomycetes: Eubacteriales
Bacillus subtilis Cohn.	Hay bacillus	Schizomycetes: Eubacteriales
Bacillus thuringiensis Berliner	Bt	Schizomycetes: Eubacteriales
Bactrocera oleae Gml. (= *Dacus oleae*)	Olive fruit fly	Diptera: Tephritidae
Basidiomycotina	Fungi, spores produced exogenously in basidia	
Beauvaria bassiana Balsamo	White muscardine	Deuteromycetes: Moniliales
Begonia elatior Hort.	Begonia	Begoniaceae
Belonolaimus longicausatus Rau	Sting nematode	Nematoda
Bemisia spp.	Whiteflies	Hemiptera: Aleyrodidae
Bemisia argentifolii Bellows & Perring	Silverleaf whitefly	Hemiptera: Aleyrodidae
Bemisia tabaci (Gennadius)	Tobacco whitefly	Hemiptera: Aleyrodidae
Betula lutea Michx.	Yellow birch	Betulaceae
Bilderdykia convolvulus Dum. (see *Fallopia convolvulus*)		
Bipolaris stenospila Shoemaker	Brown stripe, sugar cane	Ascomycetes: Sphaeriales
Blatella germanica (L.)	German cockroach	Dictyoptera: Blattidae
Blissus leucopterus (Say)	Chinch bug	Hemiptera: Lygaeidae
Blumeriella jaapii Arx.	Coccomycosis; cherry leaf spot	Ascomycetes: Helotiales
Botryosphaeria dothidea Ces. & De Not	Fruit rot, leaf spot and stem canker, apple	Ascomycetes: Sphaeriales
Botryosphaeria obtusa Shoemaker = *Physalospora obtusa* (Schw.) Cke.	Leaf spot and black rot, apple	Ascomycetes: Sphaeriales
Botrytis allii Munn	Neck rot, onion	Deuteromycetes: Moniliales
Botrytis cinerea Pers.	Fruit rot, various hosts	Deuteromycetes: Moniliales
Brachiaria mutica (Forssk.) Stapf.	Para grass	Poaceae
Bradysia spp.	Sciarid flies	Diptera: Sciaridae
Brassica napus L.	Rape	Brassicaceae
Bremia lactucae Regel	Downy mildew, lettuce	Oomycetes: Peronosporales
Brevipalpus phoenicis (Geijskes)	Red crevice tea mite	Acari: Tenuipalpidae
Bromus sterilis L.	Barren brome	Poaceae
Bryobia praetiosa Koch	Clover bryobia mite	Acari: Tetranychidae
Bryobia ribis Thom.	Gooseberry bryobia mite	Acari: Tetranychidae
Bryophyta	Mosses and liverworts	Bryophyta
Bucculatrix thurberiella Busck	Cotton leaf perforator	Lepidoptera: Lyonetiidae
Butomus umbellatus L.	Rush, flowering	Butomaceae

Latin	English	Order and/or Family
Cacopsylla pyri (L.)	Pear psylla	Hemiptera: Psyllidae
Cacopsylla pyricola (Förster)	Pear psylla	Hemiptera: Psyllidae
Calepitrimerus spp.	Mites	Acari: Eriophyidae
Caloptilia theivora (Wlsm.)	Tea leaf roller	Lepidoptera: Gracillariidae
Calystegia sepium R. Br. ssp. *Sepium*	Bindweed, large	Convolvulaceae
Camponotus spp.	Carpenter ants	Hymenoptera: Formicidae
Candida spp.	Parasitic yeasts	Endomycetales
Capsella bursa-pastoris Medic.	Shepherd's purse	Brassicaceae
Carduus spp.	Thistles	Asteraceae
Carex spp.	Sedges	Cyperaceae
Carposina niponensis Walsingham	Oriental fruit tree moth	Lepidoptera: Tortricidae
Cassia obtusifolia L.	Sickle pod	Fabaceae
Cecidomyiidae	Gall midges; predacious midges	Diptera
Cecidophyopsis ribis (Westwood)	Blackcurrant gall mite	Acari: Eriophyidae
Centaurea spp.	Knapweeds	Asteraceae
Centaurea calcitrapa L.	Purple star-thistle	Asteraceae
Centaurea cyanus L.	Cornflower	Asteraceae
Centaurea diffusa Monnet De La Marck	Diffuse knapweed	Asteraceae
Centaurea maculosa Monnet De La Marck	Spotted knapweed	Asteraceae
Centaurea solstitialis L.	Yellow star-thistle	Asteraceae
Ceratitis capitata L.	Mediterranean fruit fly	Diptera: Tephritidae
Ceratobasidium cereale Murray & Burpee	Sharp eyespot, cereals	Basidiomycetes: Tulasnellales
Ceratocystis ulmi Moreau	Dutch elm disease	Ascomycetes: Sphaeriales
Ceratodon purpureus Brid.	Moss	Bryophyta
Ceratophyllum demersum L.	Hornweed, common	Ceratophyllaceae
Cercospora spp.	Leaf spots, various	Deuteromycetes: Moniliales
Cercospora beticola Sacc.	Leaf spot, beet	Deuteromycetes: Moniliales
Cercospora zonata Winter	Cercospora leaf spot, bean	Deuteromycetes: Moniliales
Cercosporella herpotrichoides Fron. (see *Tapesia acuformis* (Nirenberg) and *T. yallendae* Wallwork & Spooner)		
Cercosporidium spp. includes *C. sojinum* (= *Cercospora sojina*)	Frog eye; leaf spot, soybean	Deuteromycetes: Moniliales
Ceutorhynchus spp.	Brassica gall weevils; brassica stem weevils	Coleoptera: Curculionidae
Ceutorhynchus assimilis (Paykull)	Cabbage seed weevil	Coleoptera: Curculionidae
Ceutorhynchus pallidactylus (Marsham)	Cabbage stem weevil	Coleoptera: Curculionidae
Ceutorhynchus pleurostigmata (Marsh.)	Turnip gall weevil	Coleoptera: Curculionidae

Latin	English	Order and/or Family
Ceutorhynchus quadridens (Panz.) (= *Ceutorhynchus pallidactylus*)		
Chaetocnema spp.	Flea beetles	Coleoptera: Chrysomelidae
Chaetocnema concinna (Marsh.)	Mangold flea beetle	Coleoptera: Chrysomelidae
Chamomilla spp.	Mayweeds (some)	Asteraceae
Chenopodium album L.	Fat hen	Chenopodiaceae
Chilo spp.	Stem borers	Lepidoptera: Pyralidae
Chilo plejadellus Zk.	Rice stem borer	Lepidoptera: Pyralidae
Chilo suppressalis (Walker)	Rice stalk borer; rice stem borer	Lepidoptera: Pyralidae
Chondrilla juncea L.	Rush skeleton weed	Asteraceae
Chondrostereum purpureum Pouzar	Silver leaf	Agaricales: Agaricaceae
Chorioptes spp.	Mange mites	Acari: Psoroptidae
Chromatomyia syngenesia (see *Phytomyza syngenesiae*)		
Chromolaena odorata King & Rob.	Siam weed	Asteraceae
Chrysanthemum cinerariaefolium Vis. (see *Tanacetum cinerariaefolium*)		
Chrysanthemum segetum L.	Corn marigold	Asteraceae
Chrysodeixis chalcites Esper	Tomato looper	Lepidoptera: Noctuidae
Chrysomelidae	Chrysomelid beetles	Coleoptera
Chrysomphalus aonidum L.	Citrus red scale	Hemiptera: Diaspididae
Chrysomphalus dictyospermi Morgan	Palm scale	Hemiptera: Diaspididae
Chrysopa carnea Stephens (see *Chrysoperla carnea*)		
Chrysoperla carnea (Stephens)	Pearly green lacewing	Neuroptera: Chrysopidae
Chrysoteuchia caliginosellus (= *Crambus caliginosellus* (Clemens))	Grass moth	Lepidoptera: Pyralidae
Cicadellidae	Leafhoppers	Hemiptera
Cirsium arvense Scop.	Thistle, creeping	Asteraceae
Cladosporium spp.	Black mould; sooty mould	Deuteromycetes: Moniliales
Cladosporium carpophilum Lev. (see *Stigmina carpophila*)		
Cladosporium fulvum Cke. (see *Fulvia fulva*)		
Clasterosporium carpophilum Aderh. (see *Stigmina carpophila*)		
Clavibacter michiganensis (Smith) Davis *et al.*	Tomato canker	Eubacteriales
Cnaphalocrocis medinalis Gn.	Rice leaf roller	Lepidoptera: Pyralidae
Cnemidocoptes spp.	Bird skin mites	Acari: Sarcoptidae
Coccidae	Scale insects	Hemiptera

Latin	English	Order and/or Family
Coccomyces hiemalis Higgins (see *Blumeriella jaapii*)		
Coccus hesperidum L.	Brown soft scale	Hemiptera: Coccidae
Coccus spp.	Scale insects	Hemiptera: Coccidae
Cochliobolus miyabeanus Drechs.	Brown spot, rice	Ascomycetes: Sphaeriales
Cochliobolus sativus Drechs.	Foot rot; root rot, cereals and grasses	Ascomycetes: Sphaeriales
Coleoptera	Beetles	Insecta
Colletotrichum spp.	Anthracnose, various root rot and leaf curl diseases	Ascomycetes: Melanconiales
Colletotrichum atramentarium Taub. (see *C. coccodes*)		
Colletotrichum coccodes Hughes	Root rot, tomato	Ascomycetes: Melanconiales
Colletotrichum coffeanum Noack (see *Glomerella cingulata*)		
Colletotrichum gloeosporioides Penz. (see *Glomorella cingulata*)		
Colletotrichum lagenarium (Pass.) Ell. & Halst.	Anthracnose, cucurbits	Ascomycetes: Melanconiales
Colletotrichum lindemuthianum Briosi & Cavara	Anthracnose, french bean	Ascomycetes: Melanconiales
Commelina spp.	Dayflower; wandering Jew	Commelinaceae
Comstockaspis perniciosus Comstock (see *Quadraspidiotus perniciosus*)		
Conopomorpha cramerella Snellen	Cocoa pod borer	Lepidoptera: Gracillariidae
Convolvulus arvensis L.	Field bindweed	Convolvulaceae
Coptotermes formosanus Shiraki	Formosan termite	Isoptera: Rhinotermitidae
Coptotermes spp.	Termites	Isoptera: Rhinotermitidae
Coquillettidea spp.	Mosquitos	Diptera: Culicideae
Corticium cerealis (see *Ceratobasidium cereale*)		
Corticium fuciforme Wakef. (see *Laetisaria fuciformis*)		
Corticium sasakii Matsu. (see *Pellicularia sasakii*)		
Corynebacterium michiganense Jens. (see *Clavibacter michiganensis*)	Leaf spot, melon	Deuteromycetes: Moniliales
Cosmopolites sordidus (Germ.)	Banana root borer; banana corm weevil	Coleoptera: Curculionidae
Costelytra zealandica (White)	New Zealand grass grub	Coleoptera: Scarabaeidae
Cotesia spp.	Lepidopteran parasitic wasp	Hymenoptera: Aphidiidae
Crambus caliginosellus (Clemens)	Grass moth	Lepidoptera: Pyralidae
Cricetus spp.	Crickets	Saltatoria: Gryllidae
Crinipellis perniciosa (Stahel) Sing	Witches' broom	Basidiomycetes: Agaricales
Cronartium ribicola Fisch.	Blackcurrant rust	Basidiomycetes: Uredinales

Latin	**English**	**Order and/or Family**
Cryphonectria parasitica (Murrill) Barr.	Chestnut blight	Ascomycetes: Nectriaceae
Cryptolaemus montrouzieri Mulsant	Mealybug predator	Coleoptera: Coccinelidae
Cryptolestes spp.	Grain beetles	Coleoptera: Cucujidae
Cryptophlebia illepida (Butler)	Koa seed worm	Lepidoptera: Tortricidae
Cryptophlebia ombrodelta (Lower)	Macadamia nut borer	Lepidoptera: Tortricidae
Ctenarytaina eucalypti Maskell	Eucalyptus psyllid	Hemiptera: Psyllidae
Cucujidae	Flour beetles	Coleoptera
Culex spp.	Mosquitos	Diptera: Culicidae
Culex fatigans Wiedemann (= *C. quinquefasciatus*)	House mosquito	Diptera: Culicidae
Culex quinquefasciatus Say. (see *C. fatigans*)		
Culicidae	Mosquitos	Diptera
Culiseta spp.	Mosquitos	Diptera: Culicideae
Curculionidae	Weevils	Coleoptera
Curvularia spp.	Leaf spot	Moniliales: Dematiaceae
Cuscuta spp.	Dodder	Convolvulaceae
Cuscuta australis (R. Br.) Engelmann	Australian dodder	Convolvulaceae
Cuscuta chinensis Lam.	Chinese dodder	Convolvulaceae
Cuscuta europaea L.	Large dodder	Convolvulaceae
Cydia caryana Fitch	Hickory shuckworm	Lepidoptera: Tortricidae
Cydia funebrana (Treitschke)	Plum fruit moth	Lepidoptera: Tortricidae
Cydia molesta (see *Grapholitha molesta*)		
Cydia nigricana Fabricius (= *Laspeyresia nigricana)*	Pea moth	Lepidoptera: Tortricidae
Cydia pomonella L.	Codling moth	Lepidoptera: Tortricidae
Cynodon spp.	Bermuda grass, star grasses	Poaceae
Cynodon dactylon Pers.	Bermuda grass	Poaceae
Cyperus spp.	Nutsedges	Cyperaceae
Cyperus brevifolius Hassk.	Kyllinga, green	Cyperaceae
Cyperus difformis L.	Umbrella plant	Cyperaceae
Cyperus esculentus L.	Yellow nutsedge	Cyperaceae
Cyperus rotundus L.	Nutgrass	Cyperaceae
Cyperus serotinus Rottbøll	Late-flowering cyperus	Cyperaceae
Dacnusa sibirica Telenga	Chrysanthemum leaf miner parasitoid	Hymenoptera: Braconidae
Dacus spp.	Fruit flies	Diptera: Tephritidae
Dacus cucurbitae Coquillet	Melon fly	Diptera: Tephritidae
Dacus oleae (Gmelin.) (see *Bactrocera oleae*)		
Datura stramonium L.	Jimson weed; thorn apple	Solanaceae

Latin	English	Order and/or Family
Decoceras spp.	e.g. field slug	Mollusca: Gastropoda
Delia spp. (= some *Hylemya* spp.)	Root flies	Diptera: Anthomyiidae
Delia brassicae (see *D. radicum*)		
Delia coarctata (Fallen)	Wheat bulb fly	Diptera: Anthomyiidae
Delia radicum (L.)	Cabbage root fly	Diptera: Anthomyiidae
Delphasus pusillus Leconte	Whitefly predatory beetle	Coleoptera: Coccinellidae
Dendroctonus frontalis Zimmerman	Southern pine beetle	Coleoptera: Scolytidae
Dendroctonus ponderosae Hopkins	Mountain pine beetle	Coleoptera: Scolytidae
Dendroctonus pseudotsuga Hopkins	Douglas fir beetle	Coleoptera: Scolytidae
Dendroctonus rufipennis (Kirby)	Spruce beetle	Coleoptera: Scolytidae
Dermolepida albohirtum (Waterhouse)	Greyback canegrub	Coleoptera: Scarabaeidae
Deuteromycetes (= Fungi Imperfecti; mitosporitic fungi)	Fungi with no known sexual stage, or asexual stages of other fungi	
Diabrotica spp.	Corn rootworms	Coleoptera: Chrysomelidae
Diabrotica undecimpunctata Barber	Corn rootworm	Coleoptera: Chrysomelidae
Dialeurodes citri (Riley & Howard)	Citrus whitefly	Hemiptera: Aleyrodidae
Diaporthales		Ascomycetes
Diaporthe spp.	Includes stem canker fungi, various hosts	Ascomycetes: Sphaeriales
Diaporthe citri Wolf	Melanosis, citrus	Ascomycetes: Sphaeriales
Diaporthe helianthi Muntanola-Cvetkovic Mihaljcevic	Leaf spot and stem canker, sunflower	Ascomycetes: Sphaeriales
Diaprepes abbreviatus (L.)	Sugar cane rootstalk borer	Coleoptera: Curculionidae
Diaspidae (and others)	Scale insects	Hemiptera
Diatraea grandiosella Dyar	Southwestern cornborer	Lepidoptera: Pyralidae
Diatraea saccharalis (Fabricius)	Maize stalk borer; sugar cane borer	Lepidoptera: Pyralidae
Didesmococcus brevipes (Cockerell)	Scale insect	Hemiptera: Coccidae
Didymella applanata Sacc.	Spur blight, cane fruit	Ascomycetes: Sphaeriales
Didymella chrysanthemi (see *Didymella ligulicola*)		
Didymella ligulicola Arx.	Ray blight, chrysanthemum	Ascomycetes: Sphaeriales
Digitaria spp.	Crabgrasses	Poaceae
Digitaria adscendens Henr. (= *D. ciliaris*)	Crabgrass, tropical	Poaceae
Digitaria ciliaris Koeler (see *Digitaria adscendens*)		
Digitaria sanguinalis Scop.	Crabgrass	Poaceae
Diplocarpon earliana Wolf	Leaf scorch, strawberry	Ascomycetes: Helotiales
Diplocarpon rosae Wolf	Blackspot, rose	Ascomycetes: Helotiales
Diplodia spp.	Stalk rots, various hosts	Deuteromycetes: Sphaeropsidales

References

Latin	English	Order and/or Family
Diplodia pseudodiplodia Fckl. (perfect stage of *Nectria galligena*)	Apple and pear canker	Deuteromycetes: Sphaeropsidales
Diplopoda	Millepedes	Myriapoda
Diprion spp.	Sawflies	Hymenoptera: Diprionidae
Diptera	Flies	Insecta
Distantiella theobroma (Dist.)	Cocoa capsid	Heteroptera: Miridae
Ditylenchus dipsaci (Kuehn)	Stem nematode	Nematoda: Tylenchidae
Dothidiales		Ascomycetes
Drechslera graminea (see *Pyrenophora graminea*)		
Drepanopeziza ribis Hoehn. (see *Pseudopeziza ribis*)		
Drosophila spp.	Small fruit flies	Diptera: Drosophilidae
Drosophilidae	Small fruit flies	Diptera
Dryocoetes confusus Swaine	Western balsam bark beetle	Coleoptera: Scolytidae
Earias spp.	Spiny bollworms	Lepidoptera: Noctuidae
Echinochloa spp.	Barnyard grasses	Poaceae
Echinochloa colonum Link	Barnyard grass, awnless	Poaceae
Echinochloa crus-galli Beauv.	Barnyard grass	Poaceae
Echinochloa oryzicola (= *E. oryzoides*)	Cockspur, rice	Poaceae
Eichhornia crassipes Solms	Water hyacinth	Pontederiaceae
Elateridae	Click beetles; wireworms	Coleoptera
Eleocharis acicularis Roem. & Schult.	Spike rush	Cyperaceae
Eleusine indica Gaertn.	Goosegrass	Poaceae
Elodea canadensis Michx.	Water weed; Canadian pondweed	Hydrocharitaceae
Elsinoe fawcettii Bitanc. & Jenkins	Scab, citrus	Ascomycetes: Myriangiales
Elymus repens (see *Elytrigia repens*)		
Elytrigia repens (L.) Desv.	Common couch; quackgrass	Poaceae
Empoasca spp.	Cotton leafhoppers	Hemiptera: Cicadellidae
Empoasca decipiens Poali	Leafhopper	Hemiptera: Cicadellidae
Empoasca fabae (Harris)	Green leafhopper	Hemiptera: Cicadellidae
Encarsia formosa Gahan	Glasshouse whitefly parasitoid	Hymenoptera: Aphelinidae
Endomycetales	Yeasts	Ascomycetes
Endothia parasitica Anders. & Anders.	Chestnut blight	Sphaeriales: Diaporthaceae
Eotetranychus spp.	Tetranychid mites	Acari: Tetranychidae
Eotetranychus sexmaculatus (Riley)	Six-spotted mite	Acari: Tetranychidae
Eotetranychus willamettei (McGregor)	Willamette mite	Acarina: Tetranychidae
Ephestia elutella (Hübner)	Warehouse moth	Lepidoptera: Pyralidae

Latin	English	Order and/or Family
Epilachna spp.	Bean beetles	Coleoptera: Coccinellidae
Epilachna varivestis (Muls.)	Mexican bean beetle	Coleoptera: Coccinellidae
Epitrimerus pyri Nalepa	Pear rust mite	Acari: Eriophyidae
Epitrix hirtipennis (Marsh)	Tobacco flea beetle	Coleoptera: Chrysomelidae
Eretmocerus sp. nr. *californicus* Howard (see *Eretmocerus eremicus*)		
Eretmocerus eremicus Rose & Zolnerowich	Whitefly parasite	Hymenoptera: Aphelinidae
Eriophyes spp.	Mite	Acari: Eriophyidae
Eriophyidae	Eriophyid mites	Acari
Eriosoma lanigerum (Hausmann)	Woolly aphid	Hemiptera: Pemphigidae
Erwinia amylovora Winsl.	Fire blight of pome fruit	Eubacteriales: Enterobacteriaceae
Erwinia carotovora Holl.	Bacterial rot, celery; basal stem rot, cucurbits; blackleg, potato	Eubacteriales: Enterobacteriaceae
Erysiphaceae		Erysiphales
Erysiphales	Powdery mildews	Ascomycetes
Erysiphe spp.	Powdery mildew, various hosts	Ascomycetes: Erysiphales
Erysiphe betae (Vanha) Weltzien	Powdery mildew, beet crops	Ascomycetes: Erysiphales
Erysiphe cichoracearum DC.	Powdery mildew, cucurbits	Ascomycetes: Erysiphales
Erysiphe graminis DC.	Powdery mildew, cereals and grasses	Ascomycetes: Erysiphales
Erythroneura elegantula (Osborn)	Grape leafhopper	Hemiptera: Cicadellidae
Eubacteriales	Cellular bacteria	
Eucosma gloriola Heinrich	Eastern pine shoot borer	Lepidoptera: Tortricidae
Eupatorium odoratum L. (= *Chromolaena odorata*)	Siam weed	Asteraceae
Euphorbia esula L.	Leafy spurge	Euphorbiaceae
Euphorbia maculata L.	Spotted spurge	Euphorbiaceae
Eupoecilia ambiguella (Hübner)	Grape berry moth	Lepidoptera: Cochylidae
Eupterycyba jucunda Herrich-Schäffer	Potato leafhopper	Hemiptera: Cicadellidae
Eutetranychus spp.	Tetranychid mites	Acari: Tetranychidae
Eutetranychus banksi (McGregor)	Texas citrus mite	Acari: Tetranychidae
Euxoa spp.	Cutworms; dart moths	Lepidoptera: Noctuidae
Exobasidium vexans Mass.	Blister blight, tea	Basidiomycetes: Exobasidiales
Fallopia convolvulus Adans.	Black bindweed	Polygonaceae
Feltiella acarisuga (Vallot)	Mite predator	Diptera: Cecidomyiidae
Fimbristylis spp.	Fringe rushes	Cyperaceae
Fomes annosus Cke.	Butt rot, conifers	Basidiomycetes: Agaricales
Formicidae	Ants	Hymenoptera

Latin	English	Order and/or Family
Frankliniella intonsa (Trybom)	Flower thrips	Thysanoptera: Thripidae
Frankliniella occidentalis (Pergande)	Western flower thrips	Thysanoptera: Thripidae
Fuchsia hybrida Voss	Fuchsia	Onagraceae
Fulvia spp.	Leaf moulds	Deuteromycetes: Hyphales
Fulvia fulva Cif.	Leaf mould, tomato	Deuteromycetes: Hyphales
Fusarium spp.	Rots, ear blights and wilts, various hosts (imperfect fungi with perfect stages in various genera)	Deuteromycetes: Moniliales
Fusarium coeruleum Sacc.	Dry rot; postharvest rot	Deuteromycetes: Moniliales
Fusarium culmorum Sacc.	Fusarium foot and root rots, various hosts	Deuteromycetes: Moniliales
Fusarium graminearum Schwabe (see *Gibberella zeae*)		Deuteromycetes: Moniliales
Fusarium moniliforme Sheldon (see *Gibberella fujikuroi*)		
Fusarium moniliforme var. *subglutinans* Wr. & Reinking	Pitch canker disease	Deuteromycetes: Moniliales
Fusarium nivale Ces. (see *Microdochium nivalis*)		
Fusarium oxysporum Schlect.	Fusarium wilt, various hosts	Deuteromycetes: Moniliales
Galendromus occidentalis (Nesbitt)	Mite predator	Acarina: Phytoseiidae
Galium aparine L.	Cleavers	Rubiaceae
Ganoderma spp.	White rot, timber	Basidiomycetes: Agaricales
Gastropoda	Slugs and snails	Mollusca
Gaultheria procumbens L.	Winterberry, teaberry	Ericacea
Geotrichum candidum Ferr.	Rubbery rot, potato	Deuteromycetes: Moniliales
Geranium spp.	Cranesbills	Geraniaceae
Gibberella spp. (= various *Fusarium* spp.)	Scab, cereals; brown foot rot and ear blight and other cereal diseases	Deuteromycetes: Hypocreales
Gibberella fujikuroi Wr.	Banana black heart; cotton boll rot; maize stalk rot; *Bakanea* disease of rice	Deuteromycetes: Hypocreales
Gibberella zeae Petch	Scab, cereals	Deuteromycetes: Hypocreales
Globodera spp.	Potato cyst nematodes	Nematoda: Heteroderidae
Globodera solanacearum Miller & Gray	Tobacco cyst nematode	Nematoda: Heteroderidae
Gloeodes pomigena Colby	Sooty blotch, apple, pear and citrus	Deuteromycetes: Sphaeropsidales
Gloeosporium spp.	Gloeosporium rot, apple	Deuteromycetes: Sphaeriales
Gloeosporium fructigenum Berk. (see *Glomerella cingulata*)		

Latin	English	Order and/or Family
Glomerella cingulata Spauld. & Schrenk	Gloeosporium rot, apple	Deuteromycetes: Sphaeriales
Gnathotricus retusus (LeConte)	Ambrosia beetle	Coleoptera: Scolytidae
Gnachotrichus sulcatus (LeConte)	Ambrosia beetle	Coleoptera: Scolytidae
Gonipterus gibberus Boisduval	Eucalyptus snout beetle	Coleoptera: Curculionidae
Grapholitha molesta (Busck)	Oriental fruit moth	Lepidoptera: Tortricidae
Gryllidae	True crickets	Orthoptera: Gryllotalpidae
Gryllotalpa spp.	Mole crickets	Orthoptera: Gryllotalpidae
Gryllotalpa gryllotalpa (L.)	Mole cricket	Orthoptera: Gryllotalpidae
Guignardia bidwellii Viala & Rivas	Black rot, grapevine	Ascomycetes: Sphaeriales
Gymnosporangium spp.	Leaf scorch, apple; rust, various hosts	Basidiomycetes: Uredinales
Gymnosporangium fuscum DC.	Pear rust	Basidiomycetes: Uredinales
Harmonia axyridis Pallas	Ladybird; ladybug	Coleoptera: Coccinellidae
Hauptidia maraccana Melichar	Leafhopper	Hemiptera: Cicadellidae
Hedera helix L.	Ivy	Araliaceae
Helianthus annuus L.	Sunflower	Asteraceae
Helicotylenchus spp.	Spiral nematodes	Nematoda: Tylenchidae
Helicoverpa armigera (Hübner)	Old World bollworm	Lepidoptera: Noctuidae
Helicoverpa assulta (Guen.)	Oriental tobacco budworm	Lepidoptera: Noctuidae
Helicoverpa zea Boddie	American bollworm	Lepidoptera: Noctuidae
Heliothis armigera (see *Helicoverpa armigera*)		
Heliothis assulta (see *Helicoverpa assulta*)		
Heliothis virescens (Fabricius)	Tobacco budworm	Lepidoptera: Noctuidae
Heliothis zea (see *Helicoverpa zea*)		
Helminthosporium oryzae B. de Haan	Helminthosporium blight, rice	Deuteromycetes: Moniliales
Helminthosporium solani Dur. & Mont.	Silver scurf, potato	Deuteromycetes: Moniliales
Helminthosporium turcicum Pass.	Northern leaf blight, maize	Deuteromycetes: Moniliales
Helotiales		Ascomycotina
Hemileia vastatrix Berk. & Br.	Coffee rust	Basidiomycetes: Uredinales
Hemiptera	Aphids, hoppers, etc.	Hemiptera
Hemitarsonemus latus (see *Polyphagotarsonemus latus*)		
Heterobasidion annosum Bref. (see *Fomes annosus*)		
Heterodera spp.	Lemon-shaped cyst nematodes	Nematoda: Heteroderidae
Heterodera cruciferae Fran.	Brassica cyst nematode	Nematoda: Heteroderidae
Heterodera glycines Ichinohe	Soybean cyst nematode	Nematoda: Heteroderidae
Heterodera goettingiana Liebs.	Pea cyst nematode	Nematoda: Heteroderidae
Heterodera schachtii Schm.	Beet cyst nematode	Nematoda: Heteroderidae

Latin	English	Order and/or Family
Heteroderidae	Cyst nematodes	Nematoda
Heteropeza pygmaea Winn.	Mushroom cecid	Diptera: Cecidomyiidae
Heteroptera	Bugs	Hemiptera
Hippodamia convergens Guerin	Ladybird; ladybug	Coleoptera: Coccinellidae
Homona spp.	Tortrix moths and leaf rollers	Lepidoptera: Tortricidae
Homona magnanima Diakonoff	Tea tortrix	Lepidoptera: Tortricidae
Hoplia philanthus (Foerster)	Welsh chafer	Coleoptera: Scarabaeidae
Hoplochelis marginalis (Fairmaire)	White grub	Coleoptera: Scarabaeidae
Hydrilla verticillata Presl.	Elodea, Florida	Hydrocharitaceae
Hylemya spp. (see *Delia* spp.)		
Hymenoptera	Ants; bees; wasps; sawflies	Insecta
Hypera postica (Gyllenhal)	Alfalfa weevil	Coleoptera: Curculionidae
Hypericum montanum L.	Pale St John's wort	Guttiferae
Hypericum perforatum L.	St John's wort	Guttiferae
Hyphales		Deuteromycotina
Hypoaspis aculeifer (Canestrini)	Fungus gnat predator	Mesostigmata: Laelapidae
Hypoaspis miles (Berlese)	Sciarid fly predator	Mesostigmata: Laelapidae
Hypocreales		Ascomycotina
Icerya purchasi Maskell	Cottony cushion scale	Hemiptera: Margarodidae
Ipomoea hederacea Jacq.	Morning glory, ivyleaf	Convolvulaceae
Ipomoea purpurea Roth.	Morning glory, tall	Convolvulaceae
Ips sexdentatus (Borner)	Six-spined ips	Coleoptera: Ipidae
Ips typographus (L.)	Douglas fir beetle	Coleoptera: Ipidae
Ischaemum rugosum Salisb.	Saramatta grass	Poaceae
Juncus maritimus Lam.	Sea-rush	Juncaceae
Jussiaea spp.	Water primroses	Onagraceae
Jussiaea diffusa (see *Ludwigia peploides*)		
Keiferia lycopersicella (Walsingham)	Tomato pinworm	Lepidoptera: Gelechiidae
Kochia scoparia Roth.	Mock cypress	Chenopodiaceae
Laetisaria fuciformis (McAlp.) Burdsall	Red thread	Basidiomycetes: Aphyllophorales
Lamium purpureum L.	Red dead-nettle	Labiatae
Laodelphax striatella (Fallen)	Small brown planthopper	Hemiptera: Delphacidae
Lapsana communis L.	Nipplewort	Asteraceae
Laspeyresia nigricana (see *Cydia nigricana*)		
Lepidoglyphus (*Glycyphagus*) *destructor* (Schrank)	Storage mite	Acarina: Acaridae
Lepidoptera	Butterflies; moths	Insecta

Latin	English	Order and/or Family
Lepidosaphes (*Cornuaspis*) *beckii* (Newmann)	Citrus mussel scale; citrus purple scale	Hemiptera: Coccidae
Lepidosaphes yangicola (Kuwana)	*Euonymus alatus* scale	Hemiptera: Coccidae
Leptinotarsa decemlineata (Say)	Colorado beetle	Coleoptera: Chrysomelidae
Leptochloa spp.	Sprangletop grasses	Poaceae
Leptochloa chinensis Nees	Sprangletop, red	Poaceae
Leptochloa fascicularis Gray (= *Diplachne fascicularis*)	Sprangletop, bearded	Poaceae
Leptomastix dactylopii Howard	Mealybug parasite	Hymenoptera: Encrytidae
Leptosphaeria nodorum Muell. (= *Septoria nodorum*)	Glume blotch, wheat	Ascomycetes: Sphaeriales
Leucoptera spp.	Leaf-mining moths	Lepidoptera: Lyonetiidae
Leucoptera malifoliella (Costa)	Pear leaf blister moth	Lepidoptera: Lyonetiidae
Leucoptera scitella (see *L. malifoliella*)		
Leveillula spp.	Powdery mildew	Ascomycetes: Erysiphales
Leveillula taurica Arn.	Tomato powdery mildew; pepper powdery mildew	Ascomycetes: Erysiphales
Lindernia procumbens Philcox	Pimpernel, false	Scrophulariaceae
Liriomyza spp.	Leaf miners	Diptera: Agromyzidae
Liriomyza bryoniae (Kalt.)	Tomato leaf miner	Diptera: Agromyzidae
Liriomyza huidobrensis (Blan.)	South American leaf miner	Diptera: Agromyzidae
Liriomyza trifolii (Burgess)	American serpentine leaf miner	Diptera: Agromyzidae
Lissorhoptrus oryzophilus Kusch	Rice water weevil	Coleoptera: Curculionidae
Lithocolletis spp. (see *Phyllonorycter* spp.)		
Lobesia botrana (Denis & Schiffermueller)	European grapevine moth	Lepidoptera: Tortricidae
Lolium spp.	Ryegrasses	Poaceae
Lolium multiflorum Lam.	Ryegrass, Italian	Poaceae
Lolium perenne L.	Ryegrass, perennial	Poaceae
Lolium rigidum Gaud.	Wimmera ryegrass; annual ryegrass	Poaceae
Longidorus spp.	Needle nematodes	Nematoda
Ludwigia palustris Ell.	Water purslane	Onagraceae
Ludwigia peploides (Kunth) Raven	Creeping water primrose	Onagraceae
Lycoriella auripila (Fitch)	Mushroom sciarid	Diptera: Sciaridae
Lygocoris pabulinus L.	Common green capsid	Heteroptera: Miridae
Lygus herperus Knight	Western tarnished plant bug	Heteroptera: Miridae
Lygus lineolaris (Palisot de Beauvois)	Southern tarnished plant bug	Heteroptera: Miridae
Lygus pabulinus L.	Tarnished plant bug	Heteroptera: Miridae
Lymantria dispar (L.)	Gypsy moth	Lepidoptera: Lymantriidae
Lyonetia clerkella (L.)	Apple leaf miner	Lepidoptera: Lyonetiidae

Latin	English	Order and/or Family
Macrosiphum euphorbiae (Thomas)	Potato aphid	Hemiptera: Aphididae
Macrosiphum rosae (L.)	Rose aphid	Hemiptera: Aphididae
Macrotermes spp.	Macrotermites	Uniramia: Macrotermitinae
Magnaporthe grisea (Hebert) Barr (see also *Pyricularia oryzae*)	Rice blast (perfect stage)	Ascomycetes: Pezizales
Magnaporthe salvinii (Cattaneo) Krause & Webster (see also *Sclerotium oryzae* Cattaneo and *Nakataea sigmoidae* (Cavara) Hara)	Rice stem rot	Ascomycetes: Pezizales
Mahanarva postica (Stål)	Sugarcane spittle bug	Hemiptera: Cercopidae
Malva spp.	Mallow	Malvaceae
Mamestra brassicae (L.)	Cabbage moth	Lepidoptera: Noctuidae
Mamestra configurata Walker	Bertha armyworm	Lepidoptera: Noctuidae
Marasmius oreades and other species	Fairy rings	Agaricales
Margarodidae	Scale insects	Hemiptera
Marsilea spp.	Four-leaved water clover	Marsileaceae
Marssonina spp.	Leaf blotches etc., various hosts	Melanconiales
Marssonina potentillae ssp. *Fragariae* (see *Diplocarpon earliana*)		
Mastigomycotina	Primitive fungi (Phycomycetes) producing motile spores	
Matricaria spp.	Mayweeds (some)	Asteraceae
Matricaria inodora (= *Tripleurospermum inodorum* Hyl. and *M. perforata* L.)	Scentless mayweed	Asteraceae
Megalurothrips sjostedti (Trybom)	Legume flower thrips	Thysanoptera: Thripidae
Megaselia spp.	Scuttle flies	Diptera: Phoridae
Melanconiales		Deuteromycotina
Meligethes spp.	Blossom beetles; pollen beetles	Coleoptera: Nitidulidae
Meligethes aeneus (Fabricius)	Pollen beetle	Coleoptera: Nitidulidae
Meloidogyne spp.	Root-knot nematodes	Nematoda
Meloidogyne incognita (K. & W.)	Southern root-knot nematode	Nematoda
Melolontha spp.	Cockchafers	Coleoptera: Scarabaeidae
Melolontha hippocastani (L.)	Cockchafer	Coleoptera: Scarabaeidae
Melolontha melolontha (L.)	Cockchafer	Coleoptera: Scarabaeidae
Metaphycus bartletti Annecke & Mynhardt	Scale insect parasite	Hymenoptera: Encrytidae
Metaphycus helvolus (Compére)	Scale insect parasite	Hymenoptera: Encrytidae
Metcalfa pruinosa Say	Citrus planthopper	Hemiptera: Fulgoroidea
Microdochium nivalis (Fr.) Samuels & Hallett	Snow mould, grasses and cereals	Deuteromycetes: Moniliales
Microtermes species	Macrotermites	Uniramia: Macrotermitinae

Latin	English	Order and/or Family
Microthyriella rubi Petr. (see *Schizothyrium pomi*)		
Miridae	Capsid bugs	Heteroptera
Mollusca	Slugs and snails	
Monilia spp.	Various rots	Deuteromycetes: Moniliales
Monilia laxa Sacc. (see *Sclerotinia fructigena*)		
Monilia roreri Cif.	Pod rot, cocoa	Deuteromycetes: Moniliales
Monilinia spp. (see *Sclerotinia* spp.)		
Monilinia laxa Honey (see *Sclerotinia fructigena*)		
Monilinia mali Honey	Monilinia leaf blight, apple	Deuteromycetes: Moniliales
Monochoria vaginalis Presl.	Pickerel weed	Pontederiaceae
Monographella nivalis	Rice leaf scald	Ascomycetes: Sphaeriales
Monomorium spp.	Seed-eating ants	Hymenoptera: Formicidae
Monomorium pharaonis (L.)	Pharaoh's ant	Hymenoptera: Formicidae
Morrenia odorata Lindl.	Strangler vine; milkweed vine	Asclepiadaceae
Mucor spp.	Fruit rot, strawberry	Mucorales
Mucorales		Zygomycotina
Musca domestica L.	Housefly	Diptera: Muscidae
Mycogone perniciosa Magn.	White mould, mushrooms	Deuteromycetes: Moniliales
Mycosphaerella spp.	Leaf spot diseases, various hosts	Ascomycetes: Sphaeriales
Mycosphaerella arachidis Deighton	Brown spot, peanut	Ascomycetes: Sphaeriales
Mycosphaerella brassicicola Dud.	Ring-spot, brassicas	Ascomycetes: Sphaeriales
Mycosphaerella fijiensis Deighton	Black leaf streak, banana	Ascomycetes: Sphaeriales
Mycosphaerella fragariae Lindau	White leaf spot, strawberry	Ascomycetes: Sphaeriales
Mycosphaerella graminicola (Fuckel) Schrot. (= *Septoria tritici* Rob.)	Septoria leaf spot, wheat	Ascomycetes: Sphaeriales
Mycosphaerella musicola Leach	Banana leaf spot; sigatoka disease	Ascomycetes: Sphaeriales
Mycosphaerella pinodes Stone (see *Ascochyta pinodes*)		
Mycosphaerella pomi Lindau	Brooks spot, apple	Ascomycetes: Sphaeriales
Myzus nicotianae Blackman	Aphid	Hemiptera: Aphididae
Myzus persicae (Sulzer)	Peach-potato aphid	Hemiptera: Aphididae
Nakataea sigmoidae (Cavara) Hara (see *Magnaporthe salvinii* (Cattaneo) Krause & Webster)		
Nectria galligena Bres. (imperfect stage of *Diplodia pseudodiplodia* Fckl.)	Apple and pear canker,	Hypocreales
Nematoda	Nematodes	

References

Latin	English	Order and/or Family
Neodiprion abietis (Harris)	Balsam whitefly	Hymenoptera: Diprionidae
Neodiprion lecontei Fitch	Pine sawfly	Hymenoptera: Diprionidae
Neodiprion sertifer Geoffrey	European pine sawfly	Hymenoptera: Diprionidae
Nephotettix spp.	Green leafhoppers	Hemiptera: Cicadellidae
Nephotettix cincticeps (Uhl.)	Green rice leafhopper	Hemiptera: Cicadellidae
Nephotettix nigropictus (Stål)	Tropical green rice leafhopper	Hemiptera: Cicadellidae
Nicotiana spp.	Tobacco	Solanaceae
Nicotiana rustica L.	Tobacco	Solanaceae
Nilaparvata spp.	Planthoppers	Hemiptera: Delphacidae
Nilaparvata lugens (Stal.)	Rice brown planthopper	Hemiptera: Delphacidae
Nitrosomonas spp.	N-fixing bacteria	Bacteria
Noctua spp.	Cutworms	Lepidoptera: Noctuidae
Noctua pronuba (L.)	Yellow underwing moth; cutworm	Lepidoptera: Noctuidae
Noctuidae	Noctuid moths	Lepidoptera
Odontotermes spp.	Macrotermites	Uniramia: Macrotermitinae
Oidiopsis taurica Salm	Tomato powdery mildew	Deuteromycetes: Moniliales
Oidium hevea Steinm.	Powdery mildew	Ascomycetes: Erysiphales
Oligonychus illicis (McGregor)	Southern red mite	Acari: Tetranychidae
Oligonychus perseae Tuttle, Baker & Abbatiello	Persea mite	Acari: Tetranychidae
Oligonychus punicae (Hirst)	Avocado brown mite	Acari: Tetranychidae
Oligonychus ununguis (Jacobi)	Spruce spider mite	Acari: Tetranychidae
Oomycetes		Mastigomycotina
Oospora lactis Sacc. (see *Geotrichum candidum*)		
Oospora pustulans Owen & Wakef. (see *Polyscytalum pustulans*)		
Opomyza spp.	Grass flies; cereal flies	Diptera: Opomyzidae
Opomyza florum (Fabricius)	Yellow cereal fly	Diptera: Opomyzidae
Opuntia lindheimerii Engelmann	Prickly pear cactus	Cactacea
Orgyia leucostigma (Sm.)	Whitemarked tussock moth	Lepidoptera: Lymantriidae
Orgyia pseudotsugata (McDunnough)	Douglas fir tussock moth	Lepidoptera: Lymantriidae
Orius spp.	Minute pirate bugs	Hemiptera: Anthocoridae
Orobanche spp.	Broomrape	Orobanchaceae
Oryctes rhinoceros (L.)	Coconut rhinoceros beetle	Coleoptera: Scarabaeidae
Oscinella frit (L.)	Frit fly	Diptera: Chloropidae
Ostrinia furnacalis Guenée	Asiatic cornborer	Lepidoptera: Pyralidae
Ostrinia nubilalis (Hübner)	European cornborer	Lepidoptera: Pyralidae
Otiorhynchus sulcatus (Fabricius)	Vine weevil	Coleoptera: Curculionidae
Oulema melanopus (L.)	Cereal leaf beetle	Coleoptera: Chrysomelidae
Oulema oryzae (Kuway.)	Rice leaf beetle	Coleoptera: Chrysomelidae

Latin	English	Order and/or Family
Pachnaeus litus Germar	Citrus weevil	Coleoptera: Curculionidae
Paecilomyces spp.	Saprophytic fungi	Deuteromycetes: Hyphales
Pandemis heparana (Denis & Schiffermueller)	Leaf roller	Lepidoptera: Tortricidae
Panicum spp.	Panic grasses	Poaceae
Panicum dichotomiflorum Michx.	Fall panicum; smooth witchgrass	Poaceae
Panicum purpurascens Raddi (see *Brachiaria mutica*)		
Panicum repens L.	Torpedo grass	Poaceae
Panicum texanum Buckl.	Texas millet	Poaceae
Panonychus spp.	Red spider mites	Acari: Tetranychidae
Panonychus citri (McGregor)	Citrus red mite	Acari: Tetranychidae
Panonychus ulmi (Koch)	Fruit tree red spider mite	Acari: Tetranychidae
Papaver spp.	Poppies	Papaveraceae
Parapediasia teterrella (Zincken)	Bluegrass webworm	Lepidoptera: Crambidae
Parlatoria ziziphi (Lucas)	Citrus black scale	Hemiptera: Diaspididae
Parthenothrips dracaenae (Hegeer)	Palm thrips	Thysanoptera: Thripidae
Pectinophora gossypiella (Saunders)	Pink bollworm	Lepidoptera: Gelechiidae
Pegomya betae (see *P. hyoscamni*)		
Pegomya hyoscamni (Panzer)	Beet leafminer; mangold fly	Diptera: Anthomyiidae
Pellicularia spp.	Rots, damping off, etc., various hosts	Basidiomycetes: Agaricales
Pellicularia sasakii Ito	Rice sheath blight	Basidiomycetes: Agaricales
Penicillium spp.	Penicillium rots	Deuteromycetes: Moniliales
Penicillium digitatum Sacc.	Green mould, citrus	Deuteromycetes: Moniliales
Penicillium expansum Link	Blue mould; blue rot, apple and pear	Deuteromycetes: Moniliales
Penicillium italicum Wehm.	Blue mould, citrus	Deuteromycetes: Moniliales
Periplaneta americana (L.)	American cockroach	Dictyoptera: Blattidae
Peronosclerospora spp.	Downy mildew, sorghum	Oomycetes: Peronosporales
Peronospora spp.	Downy mildews	Oomycetes: Peronosporales
Peronospora parasitica Fr.	Downy mildew, brassicae	Oomycetes: Peronosporales
Peronospora tabacina Adam (= *Plasmopara tabacina*)	Blue mould, tobacco	Oomycetes: Peronosporales
Peronosporales	Downy mildews, etc.	Oomycetes
Petunia spp.	Petunia	Solanaceae
Phakopsora pachyrhizi Syd.	Rust, soybean	Basidiomycetes: Uredinales
Phalaris spp.	Canary grasses	Poaceae
Phalaris paradoxa L.	Canary grass, awned	Poaceae
Phoma spp.	Root and stem rots, various	Deuteromycotina: Sphaeropsidales
Phoma exigua Desh. var. *foveata*	Gangrene, potato	Deuteromycetes: Sphaeropsidales

Latin	English	Order and/or Family
Phomopsis spp. (see *Diaporthe* spp.)		
Phomopsis citri Fawc. (see *Diaporthe citri*)		
Phomopsis helianthi (see *Diaporthe helianthi*)		
Phomopsis viticola Sacc.	Dead arm, grapevine	Deuteromycetes: Sphaeropsidales
Phoridae	Scuttle flies	Diptera
Phorodon humuli (Schrank)	Damson-hop aphid	Hemiptera: Aphididae
Phragmidium mucronatum Schlect.	Rust, rose	Basidiomycetes: Uredinales
Phragmidium violaceum (Schultz) G. Winter	Rust, blackberry	Basidiomycetes: Uredinales
Phthorimaea operculella Zeller	Potato moth	Lepidoptera: Gelechiidae
Phycomycetes	Primitive fungi with coenocytic mycelium; includes the Divisions Mastigomycotina and Zygomycotina	
Phyllactinia spp.	Powdery mildew, various hosts	Ascomycetes: Erysiphales
Phyllocoptes spp.	Mites	Acari: Eriophyidae
Phyllocoptruta spp.	Rust mites	Acari: Eriophyidae
Phyllocoptruta oleivora (Ashm.)	Citrus rust mite	Acari: Eriophyidae
Phyllonorycter spp.	Leaf-mining moths	Lepidoptera: Gracillariidae
Phyllonorycter blancardella (Fabricius)	Apple leafminer	Lepidoptera: Gracillariidae
Phyllopertha horticola (L.)	Garden chafer; bracket chafer	Coleoptera: Scarabaeidae
Phyllotreta spp.	Flea beetles	Coleoptera: Chrysomelidae
Phyllotreta striolata (Fabricius)	Flea beetle	Coleoptera: Chrysomelidae
Physalospora obtusa (Schw.) Cke. (see *Botryosphaeria obtusa*)		
Phytomyza spp.	Leaf miners	Diptera: Agromyzidae
Phytomyza syngenesiae (Hardy)	Chrysanthemum leaf miner	Diptera: Agromyzidae
Phytophthora spp.	Blight, damping off, foot rot, various hosts	Oomycetes: Peronosporales
Phytophthora cactorum Schroet.	Collar rot; crown rot, apple	Oomycetes: Peronosporales
Phytophthora capsici Leonian	Blight, capsicums	Oomycetes: Peronosporales
Phytophthora fragariae Hickman	Red core, strawberry	Oomycetes: Peronosporales
Phytophthora infestans De Bary	Late blight, potato and tomato	Oomycetes: Peronosporales
Phytophthora megasperma Drechs.	Root rot, brassicas	Oomycetes: Peronosporales
Phytophthora palmivora (Butl.) Butl.	Rot, various crops	Oomycetes: Peronosporales
Phytoseiulus persimilis Athios-Henriot	Two-spotted spider mite predator	Mesostigmata: Phytoseiidae
Pieris spp.	Cabbage white butterflies	Lepidoptera: Pieridae
Pieris brassicae (L.)	Large white butterfly	Lepidoptera: Pieridae
Pieris rapae (L.)	Small white butterfly	Lepidoptera: Pieridae

Latin	English	Order and/or Family
Pikonema alaskensis (Rohwer)	Yellow-headed spruce sawfly	Hymenoptera: Tenthredinidae
Pistia stratiotes L.	Water duckweed	Araceae
Pityogenes chalcographus (L.)	Six-toothed spruce bark beetle	Coleoptera: Scolytidae
Planococcus citri (Risso)	Citrus mealybug	Hemiptera: Pseudococcidae
Plantago spp.	Plantains	Plantaginaceae
Plasmodiophora brassicae Woron.	Clubroot, brassicas	Phycomycetes: Plasmodiophorales
Plasmodiophoromycetes	Parasitic members of the Myxomycota – a group with affinities with both primitive fungi and primitive animals	
Plasmopara spp.	Downy mildews, various hosts	Oomycetes: Peronosporales
Plasmopara tabacina (see *Peronospora tabacina*)		
Plasmopara viticola Berl. & de T.	Downy mildew, grapevine	Oomycetes: Peronosporales
Platynota idaeusalis (Walker)	Tufted apple moth	Lepidoptera: Tortricidae
Platynota stultana (Walsingham)	Leaf roller	Lepidoptera: Tortricidae
Platyptilia carduidactyla (Riley)	Artichoke plume moth	Lepidoptera: Pterophoridae
Plodia interpunctella (Hübner)	Indian meal moth	Lepidoptera: Pyralidae
Plusia spp.	e.g. silvery moth	Lepidoptera: Noctuidae
Plutella xylostella (L.)	Diamondback moth	Lepidoptera: Yponomeutidae
Poa spp.	Meadowgrasses	Poaceae
Poa annua L.	Meadowgrass, annual	Poaceae
Poa trivialis L.	Meadowgrass, rough	Poaceae
Podisus maculiventris (Say)	Caterpillar predator	Heteroptera: Pentatomidae
Podosphaera spp.	Powdery mildew, various hosts	Ascomycetes: Erysiphales
Podosphaera leucotricha (Ellis & Everhart) Salmon	Powdery mildew, apple	Ascomycetes: Erysiphales
Polychrosis botrana (see *Lobesia botrana*)		
Polygonum spp.	Knotweeds	Polygonaceae
Polygonum aviculare L.	Knot grass	Polygonaceae
Polygonum convolvulus L. (see *Fallopia convolvulus*)		
Polygonum cuspidatum Sieb. & Zucc. (see *Reynoutria japonica*)		
Polygonum lapathifolium L.	Pale persicaria	Polygonaceae
Polygonum persicaria L.	Redshank; persicaria; smartweed	Polygonaceae
Polygonum sachalinense Schmidt (see *Reynoutria sachalinensis*)		
Polymyxa betae Keskin	Fungal vector of rhizomania virus	Phycomycetes: Plasmodiophorales

Latin	English	Order and/or Family
Polyphagotarsonemus latus (Banks)	Broad mite	Acari: Tarsonemidae
Polyscytalum pustulans (Owen & Wakef.) Ellis	Skin spot, potato	Deuteromycetes: Hyphales
Polystigmatales		Ascomycetales
Popillia japonica Newman	Japanese beetle	Coleoptera: Scarabaeidae
Populus spp.	Poplars	Salicaceae
Portulaca spp.	Purslanes	Portulacaceae
Portulaca oleracea L.	Purslane	Portulacaceae
Potamogeton spp.	Pondweeds	Potamogetonaceae
Potamogeton distinctus Benn.	Pondweed, American	Potamogetonaceae
Pratylenchus spp.	Root-lesion nematodes	Nematoda: Hoplolaimidae
Prays citri Millière	Citrus flower moth	Lepidoptera: Yponomeutidae
Prays oleae (Bernard)	Olive moth	Lepidoptera: Yponomeutidae
Prunus serotina Ehrh.	American black cherry	Rosaceae
Pseudocercosporella capsellae (Ell. & Ev.)	White leaf spot, oilseed rape	Deuteromycetes: Hyphales
Pseudocercosporella herpotrichoides Deighton (see *Tapesia acuformis* (Nirenberg) and *T. yallendae* Wallwork & Spooner)		
Pseudococcidae	Mealybugs	Hemiptera
Pseudococcus spp.	Mealybugs	Hemiptera: Pseudococcidae
Pseudococcus affinis (Maskell)	Obscure mealybug	Hemiptera: Pseudococcidae
Pseudococcus longispinus (Targioni-Tozzetti)	Long-tailed mealybug	Hemiptera: Pseudococcidae
Pseudomonas spp.	Bacterial blights and leaf spots, various hosts	Pseudomonadales: Pseudomonadaceae
Pseudomonas glumae Kurita & Tabei	Bacterial grain rot, rice	Pseudomonadales: Pseudomonadaceae
Pseudomonas lachrymans Carsner	Angular leaf spot, cucurbits	Pseudomonadales: Pseudomonadaceae
Pseudomonas mors-prunorum Wormold	Bacterial canker, prunus	Pseudomonadales: Pseudomonadaceae
Pseudomonas phaseolicola Dows	Halo blight, bean	Pseudomonadales: Pseudomonadaceae
Pseudomonas solanacearum (Smith) Smith (see also *Ralstonia solanacearum* (Smith))	Bacterial wilt; brown rot	Pseudomonadales: Pseudomonadaceae
Pseudomonas syringae Van Hall pv. *lachrymans* (see *P. lachrymans*)		Pseudomonadales: Pseudomonadaceae
Pseudomonas syringae Van Hall pv. *mors-prunorum* (see *P. mors-prunorum*)		Pseudomonadales: Pseudomonadaceae

Latin	English	Order and/or Family
Pseudomonas tabaci Stevens	Wild fire of tobacco and soybean	Pseudomonadales: Pseudomonadaceae
Pseudomonas tolaasii Paine	Brown fleck of mushrooms	Pseudomonadales: Pseudomonadaceae
Pseudoperonospora cubensis Rostow	Downy mildew, cucurbits	Oomycetes: Peronosporaceae
Pseudoperonospora humuli Wils.	Downy mildew, hop	Oomycetes: Peronosporaceae
Pseudopeziza ribis Kleb.	Leaf spot, currant and gooseberry	Ascomycetes: Helotiales
Pseudoplusia includens (Walker)	Soybean looper	Lepidoptera: Noctuidae
Pseudotsuga menziesii Franco	Douglas fir	Pinaceae
Psila rosae (Fabricius)	Carrot fly	Diptera: Psilidae
Psylla spp.	Psyllids	Hemiptera: Psyllidae
Psyllidae	Psyllids	Hemiptera
Pteridium aquilinum Kuhn	Bracken	Filicales
Puccinia spp.	Rust, various hosts	Basidiomycetes: Uredinales
Puccinia chrysanthemi Roze	Brown rust, chrysanthemum	Basidiomycetes: Uredinales
Puccinia graminis Pers.	Black stem rust, grasses	Basidiomycetes: Uredinales
Puccinia hordei Otth.	Brown rust, barley	Basidiomycetes: Uredinales
Puccinia recondita Rob. (see *Puccinia triticina)*		Basidiomycetes: Uredinales
Puccinia striiformis West	Yellow rust, cereals	Basidiomycetes: Uredinales
Puccinia triticina Eriks.	Brown rust, wheat	Basidiomycetes: Uredinales
Pyralidae	Pyralid moths	Lepidoptera
Pyrenopeziza brassicae Sutton & Rawl.	Light leaf spot, brassicas	Ascomycetes: Helotiales
Pyrenophora graminea Ito & Kuribay	Leaf stripe, barley	Ascomycetes: Sphaeriales
Pyrenophora teres Drechs.	Net blotch, barley	Ascomycetes: Sphaeriales
Pyrenophora tritici-vulgaris Dicks	Tan spot, wheat	Ascomycetes: Sphaeriales
Pyrethrum cinerariaefolium Trev. (see *Tanacetum cinerariaefolium*)		
Pyricularia oryzae Cavara (see also *Magnaporthe grisea*)	Rice blast (imperfect stage)	Deuteromycetes: Moniliales
Pythium spp.	Root rots, various	Oomycetes: Peronosporales
Quadraspidiotus perniciosus (Comstock)	San José scale	Hemiptera: Coccidae
Radopholus similis (Cobb) Thorne	Burrowing nematode	Nematoda: Tylenchidae
Ralstonia solanacearum (Smith) (see also *Pseudomonas solanacearum* (Smith) Smith)	Bacterial wilt; brown rot	Burkholderiales: Burkholderiaceae
Ramularia spp.	Leaf spots, various	Deuteromycetes: Moniliales

References

Latin	English	Order and/or Family
Ramularia beticola Fautr. & Lambotte	Leaf spot, beet	Deuteromycetes: Moniliales
Ranunculus spp.	Buttercups	Ranunculaceae
Raphanus raphanistrum L.	Wild radish; runch	Brassicaceae
Reynoutria japonica Houtt. (= *Polygonum cuspidatum* Sieb. & Zucc.)	Japanese knotweed	Polygonaceae
Reynoutria sachalinensis Schmidt (Nakai) (= *Polygonum sachalinense*)	Giant knotweed	Polygonaceae
Rhinotermitidae	Termites	Isoptera: Rhinotermitidae
Rhizoctonia spp.	Foot rot; root rot, various hosts	Deuteromycetes: Agonomycetiales
Rhizoctonia solani Kuehn (= *Thanetophorus cucumeris*)	Damping off; root rots, various hosts	Deuteromycetes: Agonomycetiales
Rhizoglyphus callae, R. robini Claperède	Bulb mites	Acari: Acaridae
Rhizoglyphus echinopus (see *R. callae, R. robini*)		
Rhizopertha dominica (Fabricius) (see *Rhyzopertha dominica*)		
Rhizophora mangle L.	Mangrove	Rhizophoraceae
Rhizopus spp.	Postharvest rots	Phycomycetes: Mucorales
Rhododendron ponticum L.	Rhododendron	Ericaceae
Rhopalosiphum padi (L.)	Bird-cherry aphid; apple grain aphid	Hemiptera: Aphididae
Rhus aromatica Aiton	Sumac	Anacardiacea
Rhyacionia buoliana (Denis & Schiffermueller)	European pine shoot moth	Lepidoptera: Tortricidae
Rhynchophorus ferrugineus (Olivier)	Red palm weevil	Coleoptera: Curculionidae
Rhynchophorus palmarum (L.)	American palm beetle	Coleoptera: Curculionidae
Rhynchosporium spp.	Leaf spots, grasses	Deuteromycetes: Moniliales
Rhynchosporium secalis Davis	Leaf blotch, barley and rye	Deuteromycetes: Moniliales
Rhyzopertha dominica Rd.	Lesser grain borer	Coleoptora: Bostrichidae
Richia albicosta (Smith)	Western bean cutworm	Lepidoptera: Noctuidae
Rotylenchulus reniformis Linford & Oliveira	Reniform nematode	Nematoda
Rubus spp.	Brambles; blackberry	Rosaceae
Rubus fruticosus L.	Brambles; blackberry	Rosaceae
Rumex spp.	Docks and sorrels	Polygonaceae
Rumex crispus L.	Curly dock	Polygonaceae
Sagittaria sagittifolia L.	Arrowhead	Alismataceae
Sahlbergella singularis Hagl.	Cocoa capsid	Heteroptera: Miridae
Saissetia coffeae (Walker)	Hemispherical scale; helmet scale	Hemiptera: Coccidae

Latin	English	Order and/or Family
Saissetia oleae (Bernard)	Mediterranean black scale; black olive scale	Hemiptera: Coccidae
Salsola kali L.	Russian thistle	Chenopodiaceae
Saltatoria	Crickets, grasshoppers, etc.	Insecta
Sanninoidea exitiosa (Say.)	Peach tree borer	Lepidoptera: Aegeriidae
Saprolegniales		Oomycetes
Sarothamnus scoparius Wimmer	Scotchbroom	Fabaceae
Scapteriscus vicinus Scudder	Mole cricket	Orthoptera: Gryllotapidae
Scerophthora spp.	Downy mildew, wheat	Oomycetes: Peronosporales
Scerophthora macrospora	Downy mildew, cereals	Oomycetes: Peronosporales
Schizothyrium pomi Arx.	Fly speck disease, apple	Ascomycetes: Hemisphaeriales
Sciara spp.	Sciariad flies	Diptera: Sciaridae
Sciaridae	Fungus gnats; sciarid flies	Diptera
Scirpus spp.	Club-rushes	Cyperaceae
Scirpus juncoides	Japanese bullrush	Cyperaceae
Scirpus maritimus L.	Sea club-rush	Cyperaceae
Scirpus mucronatus L.	Roughseed bullrush	Cyperaceae
Scirtothrips persea Nakahara	Avocado thrips	Thysanoptera: Thripidae
Sclerospora spp.	Downy mildews, e.g. on pearl millet	Oomycetes: Peronosporales
Sclerotinia spp.	Sclerotinia rots, various hosts	Ascomycetes: Helotiales
Sclerotinia fructicola Rehm	Brown rot, top fruit	Ascomycetes: Helotiales
Sclerotinia fructigena Anderh. & Ruhl.	Brown rot, apple, pear and plum	Ascomycetes: Helotiales
Sclerotinia homeocarpa Bennett	Dollar spot, turf	Ascomycetes: Helotiales
Sclerotinia laxa Anderh. & Ruhl.	Blossom wilt, apple and plum	Ascomycetes: Helotiales
Sclerotinia minor Jagger	Lettuce neck rot	Ascomycetes: Helotiales
Sclerotinia sclerotiorum De Bary	Rots of stems, storage organs, etc., various crops	Ascomycetes: Helotiales
Sclerotium spp.	Postharvest rots, various hosts	Deuteromycetes: Agonomycetales
Sclerotium cepivorum Berk.	White rot, onion	Deuteromycetes: Agonomycetales
Sclerotium oryzae Cattaneo see *Magnaporthe salvinii* (Cattaneo) Krause & Webster		
Sclerotium rolfsii Sacc.	Rots, various hosts	Deuteromycetes: Agonomycetales
Scolytus multistratus (Marsham)	Smaller European elm bark beetle	Coleoptera: Scolytidae
Senecio jacobaea L.	Common ragwort; tansy ragwort	Asteraceae
Septoria spp.	Leaf and glume spots, various hosts	Ascomycetes: Sphaeriales

Latin	**English**	**Order and/or Family**
Septoria nodorum Berk. (see *Leptosphaeria nodorum*)		
Septoria tritici Rob. (see *Mycosphaerella graminicola*)		
Sesamia cretica Lederer	Pink cornborer	Lepidoptera: Pyralidae
Sesamia nonagroides (Lefèbvre)	Mediterranean cornborer	Lepidoptera: Pyralidae
Sesbania exaltata Cory	Hemp sesbania	Fabaceae
Setaria spp.	Foxtail grasses	Poaceae
Setaria faberi Herrm.	Foxtail, giant	Poaceae
Setaria glauca Beauv.	Foxtail, yellow	Poaceae
(= *S. lutescens* Hurb.)		
Setaria viridis Beav.	Foxtail, green	Poaceae
Sida spinosa L.	Spiny sida; prickly sida	Malvaceae
Simuliidae	Blackflies	Diptera
Simulium spp.	Blackflies	Diptera: Simuliidae
Sinapis alba L.	White mustard	Brassicaceae
Sinapis arvensis L.	Charlock	Brassicaceae
Sitobium avenae (Fabricius)	Grain aphid	Hemiptera: Aphididae
Sitona spp.	Pea weevil; bean weevil	Coleoptera: Curculionidae
Sitophilus oryzae (L.)	Rice weevil	Coleoptera: Curculionidae
Sitophilus zeamais Mutsch.	Maize weevil; rice weevil	Coleoptera: Curculionidae
Sitotroga cerealella (Oliver)	Angoumois grain moth	Lepidoptera: Gelechiidae
Sogatella furcifera (Howorth)	White-backed planthopper	Hemiptera: Delphacidae
Solanum nigrum L.	Black nightshade	Solanaceae
Solenopsis spp.	Fire ants	Hymenoptera: Formicidae
Sonchus oleraceus L.	Smooth sowthistle	Asteraceae
Sorghum spp.	Sorghum grasses	Poaceae
Sorghum almum Parodi	Columbus grass	Poaceae
Sorghum bicolor Moench	Shattercane	Poaceae
Sorghum halepense Pers.	Johnson grass	Poaceae
Sparganium erectum L.	Branched bur-reed	Sparganiaceae
Spergula arvensis L.	Corn spurrey	Caryophyllaceae
Sphacelotheca reiliana Clint.	Head smut, maize	Basidiomycetes: Ustilaginales
Spheariales		Ascomycotina
Sphaeropsidales		Deuteromycotina
Sphaerotheca spp.	Powdery mildew, various hosts	Ascomycetes: Erysiphales
Sphaerotheca fuliginea Poll.	Powdery mildew, cucurbits	Ascomycetes: Erysiphales
Sphaerotheca pannosa Lev.	Powdery mildew, rose	Ascomycetes: Erysiphales
Spodoptera spp.	Armyworms	Lepidoptera: Noctuidae
Spodoptera eridania (Cram.)	Southern armyworm	Lepidoptera: Noctuidae
Spodoptera exigua (Hübner)	Beet armyworm; lesser armyworm	Lepidoptera: Noctuidae
Spodoptera frugiperda (J. E. Smith)	Fall armyworm	Lepidoptera: Noctuidae

Latin	English	Order and/or Family
Spodoptera littoralis (Boisch.)	Egyptian cotton leafworm	Lepidoptera: Noctuidae
Spodoptera litura (Hübner)	Beet armyworm	Lepidoptera: Noctuidae
Stellaria media Vill.	Common chickweed	Caryophyllaceae
Steneotarsonemus laticeps (Halb.)	Bulb scale mite	Acari: Tarsonemidae
Stethorus punctillum (Weise)	Minute black ladybird	Coleoptera: Coccinellidae
Stigmina carpophila Ell.	Shothole, prunus	Deuteromycetes: Moniliales
Streptomyces scabies Walk.	Common scab of crops such as potato and beet	Schizomycetes: Actinomycetales
Striacosta albicosta Smith (see *Richia albicosta* Smith)		
Symphyla spp.	Symphilids	Myriapoda
Syngrapha falcifera	Celery looper	Lepidoptera; Noctuidae
Synanthedon hector (Butler)	Cherry tree borer	Lepidoptera: Aegeriidae
Synanthedon myopaeformis (Borkhausen)	Apple clearwing moth	Lepidoptera: Aegeriidae
Synanthedon pictipes (Grote & Robinson)	Lesser peach tree borer	Lepidoptera: Aegeriidae
Synanthedon tipuliformis (Clerck)	Currant clearwing moth	Lepidoptera: Aegeriidae
Tagetes erecta L.	African marigold	Asteraceae
Tanacetum cinerariaefolium (Trev.) Schultz-Bip (syns. *Chrysanthemum cinerariaefolium* Vis. and *Pyrethrum cinerariaefolium* Trev.)	Pyrethrum daisy	Asteraceae
Tanymecus palliatus Fabricius	Beet leaf weevil	Coleoptera: Curculionidae
Tapesia acuformis (Nirenberg)	Eye-spot, cereals	Deuteromycetes: Moniliales
Tapesia yallendae Wallwork & Spooner	Eye-spot, cereals	Deuteromycetes: Moniliales
Taphrina deformans Tul.	Peach leaf-curl	Ascomycetes: Taphrinales
Tarsonemus spp. (= *Phytonemus*, in part)	Tarsonemid mites	Acari: Tarsonemidae
Taxus baccata L.	Yew	Taxaceae
Tephritidae	Large fruit flies	Diptera
Tetanops myopaeformis (Roeder)	Sugar beet root maggot	Diptera: Otitidae
Tetranychidae	Spider mites	Acari
Tetranychus spp.	Spider mites	Acari: Tetranychidae
Tetranychus atlanticus McGregor	Strawberry mite	Acari: Tetranychidae
Tetranychus cinnabarinus (Boisduval)	Carmine spider mite	Acari: Tetranychidae
Tetranychus kanzawi Kishida	Kanzawa spider mite	Acari: Tetranychidae
Tetranychus mcdanieli McG.	McDaniel's spider mite	Acari: Tetranychidae
Tetranychus pacificus McG.	Pacific spider mite	Acari: Tetranychidae
Tetranychus urticae Koch	Two-spotted spider mite	Acari: Tetranychidae
Thanetophorus cucumeris Donk. (= *Rhizoctonia solani* Kuehn)	Damping-off disease	Basidiomycetes: Stereales

Latin	English	Order and/or Family
Thaumetopoea pityocampa (Denis & Schiffermueller)	Pine processionary caterpillar	Lepidoptera: Thaumetopoeidae
Thielaviopsis basicola Ferr.	Black root rot, tobacco	Deuteromycetes: Moniliales
Thripidae	Thrips	Thysanoptera
Thrips spp.	Thrips	Thysanoptera: Thripidae
Thrips fuscipennis Haliday	Rose thrips	Thysanoptera: Thripidae
Thrips obscuratus Crawford	New Zealand flower thrips	Thysanoptera: Thripidae
Thrips palmi Karny	Thrips	Thysanoptera: Thripidae
Thrips tabaci Lindeman	Onion thrips	Thysanoptera: Thripidae
Thymus vulgaris L.	Thyme	Labiatae
Thysanoptera	Thrips	Insecta
Tilletia spp.	Smut, various hosts	Basidiomycetes: Ustilaginales
Tilletia caries Tul.	Bunt; stinking smut	Basidiomycetes: Ustilaginales
Tipula spp.	Crane flies; leatherjackets	Diptera: Tipulidae
Tortricidae	Tortrix moths	Lepidoptera
Tortrix spp.	Tortrix moths	Lepidoptera: Tortricidae
Tranzschelia discolor Tranz. & Litw. (see *Tranzchelia pruni-spinosae*)		
Tranzchelia pruni-spinosae Diet.	Plum rust	Basidiomycetes: Uredinales
Trialeurodes vaporariorum (Westwood)	Glasshouse whitefly	Hemiptera: Aleyrodidae
Tribulus terrestris L.	Puncture vine	Zygophyllaceae
Trichodorus spp.	Stubby-root nematodes	Nematoda
Trichogramma spp.	Lepidopteran egg parasites	Hymenoptera: Trichogrammatidae
Trichoplusia ni (Hübner)	Cabbage looper	Lepidoptera: Noctuidae
Tripleurospermum maritimum Koch (= *Matricaria inodora* L.)	Scentless mayweed	Asteraceae
Trypodendron lineatum (Olivier)	Ambrosia beetle	Coleoptera: Scolytidae
Tulasnellales		Basidiomycotina
Typha spp.	Bullrushes	Typhaceae
Typhlodromus occidentalis Nesbitt (synonymous with *Galendromus occidentalis* & *Metaseiulus occidentalis*)	Predatory mite	Mesostigmata: Phytoseiidae
Typhlodromus pyri Scheuten	Fruit tree red spider mite predator	Mesostigmata: Phytoseiidae
Typhula incarnata Lasch	Snow rot, cereals	Basidiomycetes: Agaricales
Ulex europaeus L.	Gorse	Leguminacea
Unaspis euonymi	Euonymus scale	Hemiptera: Coccidae
Uncinula necator Burr	Powdery mildew, grapevine	Erysiphales: Erysiphaceae
Uredinales	Rust fungi	Basidiomycetes
Urocystis spp.	Leaf smuts, various hosts	Basidiomycetes: Ustilaginales

Latin	English	Order and/or Family
Uromyces spp.	Rusts, various crops	Basidiomycetes: Uredinales
Uromyces betae Lev.	Rust, beet crops	Basidiomycetes: Uredinales
Urtica spp.	Nettles	Urticaceae
Urtica dioica L.	Nettle, common	Urticaceae
Urtica urens L.	Nettle, small	Urticaceae
Ustilaginales	Smut fungi	Basidiomycetes
Ustilago spp.	Smut diseases, various hosts	Basidiomycetes: Ustilaginales
Ustilago nuda Rostr.	Loose smut, barley and wheat	Basidiomycetes: Ustilaginales
Valsa ceratosperma (Tode) Maire	Valsa canker of apple	Ascomycetes: Sphaeriales
Vasates spp.	Mites	Acari: Eriophyidae
Venturia inaequalis Wint.	Scab, apple	Ascomycetes: Sphaeriales
Venturia pirina Aderh.	Scab, pear	Ascomycetes: Sphaeriales
Veronica spp.	Speedwells	Scrophulariaceae
Veronica filiformis Sm.	Speedwell, slender	Scrophulariaceae
Veronica hederifolia L.	Speedwell, ivy-leaved	Scrophulariaceae
Veronica persica Poir.	Speedwell, common field	Scrophulariaceae
Verticillium spp.	Verticillium wilt, various hosts	Deuteromycetes: Moniliales
Verticillium albo-atrum Reinke & Berth.	Wilt	Deuteromycetes: Moniliales
Verticillium dahliae Kleb.	Black heart, apricots	Deuteromycetes: Moniliales
Verticillium fungicola (Preuss) Hassebrauk	Dry bubble, mushrooms	Deuteromycetes: Moniliales
Viola spp.	Wild pansies	Violaceae
Viola arvensis Murr.	Field pansy	Violaceae
Xanthium pennsylvanicum Wallr.	Cocklebur	Asteraceae
Xanthium strumarium L.	Rough cocklebur	Asteraceae
Xanthomonas spp.	Bacterial leaf spots, various hosts	Pseudomonadales: Pseudomonadaceae
Xanthomonas campestris Dows. pv. *citri* (see *X. citri*)		Pseudomonadales: Pseudomonadaceae
Xanthomonas campestris Dows. pv. *malvacearum* (see *X. malvacearum*)		Pseudomonadales: Pseudomonadaceae
Xanthomonas campestris Dows. pv. *oryzae* (see *X. oryzae*)		Pseudomonadales: Pseudomonadaceae
Xanthomonas citri Dows.	Citrus canker	Pseudomonadales: Pseudomonadaceae
Xanthomonas malvacearum Dows.	Bacteriosis, cotton	Pseudomonadales: Pseudomonadaceae
Xanthomonas oryzae Dows.	Leaf blight, rice	Pseudomonadales: Pseudomonadaceae
Zeuzera pyrina (L.)	Leopard moth	Lepidoptera: Cossidae

Latin	English	Order and/or Family
Zygomycotina	Primitive fungi (Phycomycetes) which do not produce motile spores; sexually produced spores are non-motile zygospores	
Zygophiala jamaicensis Mason	Greasy blotch, carnation	Deuteromycetes: Moniliales

Glossary: English–Latin

This glossary is intended to help the reader to identify the Latin name in cases where the English name may be unfamiliar. The Latin name is either presented as genus and species or, if the English name is represented by a genus, the specific name is excluded. For example, the causal organisms for anthracnose diseases are fungal pathogens from the genus *Colletotrichum* but the peach-potato aphid is *Myzus persicae*.

This glossary has been produced simply by inverting the Latin–English glossary, with limited subsequent editing. The reader should recognise that English-name and Latin-name groups of species are not congruent. For example, all Peronospora are downy mildews, but not all downy mildews are Peronospora; consequently, there are entries for various downy mildews giving different Latin names.

In searching for English names, alternative forms of the name should be considered, especially with names containing an adjectival component.

Where these two problems occur together, this glossary needs to be used with particular care.

English	**Latin**	**Order and/or Family**
Acarid mites	Acaridae	Acari
African marigold	*Tagetes erecta*	Asteraceae
Alfalfa aphid	*Acyrthosiphon kondoi*	Hemiptera: Aphididae
Alfalfa looper	*Anagrapha falcifera*	Lepidoptera: Noctuidae
Alfalfa looper	*Autographa californica*	Lepidoptera: Noctuidae
Alfalfa weevil	*Hypera postica*	Coleoptera: Curculionidae
Alligatorweed	*Alternanthera philoxeroides*	Amaranthaceae
Amaranths	*Amaranthus*	Amaranthaceae
Ambrosia beetle	*Gnathotricus retusus*	Coleoptera: Scolytidae
Ambrosia beetle	*Gnathotrichus sulcatus*	Coleoptera: Scolytidae
Ambrosia beetle	*Trypodendron lineatum*	Coleoptera: Scolytidae
American black cherry	*Prunus serotina*	Rosaceae
American bollworm	*Helicoverpa zea*	Lepidoptera: Noctuidae
American cockroach	*Periplaneta americana*	Dictyoptera: Blattidae
American palm beetle	*Rhyncophorus palmatum*	Coleoptera: Curculionidae
American serpentine leaf miner	*Liriomyza trifolii*	Diptera: Agromyzidae
Angoumois grain moth	*Sitotroga cerealella*	Lepidoptera: Gelechiidae
Angular leaf spot, cucurbits	*Pseudomonas lachrymans*	Pseudomonadales: Pseudomonadaceae

English	Latin	Order and/or Family
Annual ryegrass	*Lolium rigidum*	Poaceae
Anthracnose, cucurbits	*Colletotrichum lagenarium*	Ascomycetes: Melanconiales
Anthracnose, french bean	*Colletotrichum lindemuthianum*	Ascomycetes: Melanconiales
Anthracnose, various root rot and leaf curl diseases	*Colletotrichum*	Ascomycetes: Melanconiales
Ants	Formicidae	Hymenoptera
Aphid	*Myzus nicotianae*	Hemiptera: Aphididae
Aphid gall midge	*Aphidoletes aphidimyza*	Diptera: Cecidomyiidae
Aphid parasitoid wasps	*Aphelinus*	Hymenoptera: Aphelinidae
Aphid parasitoid wasps	*Aphidius*	Hymenoptera: Aphelinidae
Aphids	Hemiptera	Hemiptera
Apple and pear canker	*Diplodia pseudodiplodia*	Deuteromycetes: Sphaeropsidales
Apple and pear canker	*Nectria galligena*	Deuteromycetes: Hypocreales
Apple clearwing moth	*Synanthedon myopaeformis*	Lepidoptera: Aegeriidae
Apple fruit rot, leaf spot and stem canker	*Botryosphaera dothidea*	Ascomycetes: Sphaeriales
Apple grain aphid	*Rhopalosiphum padi*	Hemiptera: Aphididae
Apple leaf miner	*Phyllonorycter blancardella*	Lepidoptera: Gracillariidae
Apple leaf miner	*Lyonetia clerkella*	Lepidoptera: Lyonetiidae
Armyworm, beet	*Spodoptera exigua*	Lepidoptera: Noctuidae
Armyworm, beet	*Spodoptera litura*	Lepidoptera: Noctuidae
Armyworm, bertha	*Mamestra configurata*	Lepidoptera: Noctuidae
Armyworm, fall	*Spodoptera frugiperda*	Lepidoptera: Noctuidae
Armyworm, lesser	*Spodoptera exigua*	Lepidoptera: Noctuidae
Armyworm, southern	*Spodoptera eridania*	Lepidoptera: Noctuidae
Armyworms	*Spodoptera* spp.	Lepidoptera: Noctuidae
Arrowhead	*Sagittaria sagittifolia*	Alismataceae
Artichoke plume moth	*Platyptilia carduidactyla*	Lepidoptera: Pyralidae
Asiatic cornborer	*Ostrinia furnacalis*	Lepidoptera: Pyralidae
Avocado thrips	*Scirtothrips persea*	Thysanoptera: Thripidae
Bacterial blights and leaf spots, various hosts	*Pseudomonas*	Pseudomonadales: Pseudomonadaceae
Bacterial canker, prunus	*Pseudomonas mors-prunorum*	Pseudomonadales: Pseudomonadaceae
Bacterial grain rot, rice	*Pseudomonas glumae*	Pseudomonadales: Pseudomonadaceae
Bacterial leaf spots, various hosts	*Xanthomonas*	Pseudomonadales: Pseudomonadaceae
Bacterial rot, celery	*Erwinia carotovora*	Eubacteriales: Enterobacteriaceae
Bacterial wilt	*Pseudomonas solanacearum*	Pseudomonadales: Pseudomonadaceae

English	Latin	Order and/or Family
Bacteriosis, cotton	*Xanthomonas malvacearum*	Pseudomonadales: Pseudomonadaceae
Balsam whitefly	*Neodiprion abietis*	Hymenoptera: Diprionidae
Banana black heart	*Gibberella fujikuroi*	Deuteromycetes: Hypocreales
Banana corm weevil	*Cosmopolites sordidus*	Coleoptera: Curculionidae
Banana leaf spot	*Mycosphaerella musicola*	Ascomycetes: Sphaeriales
Banana root borer	*Cosmopolites sordidus*	Coleoptera: Curculionidae
Barnyard grass	*Echinochloa crus-galli*	Poaceae
Barnyard grass, awnless	*Echinochloa colonum*	Poaceae
Barnyard grasses	*Echinochloa*	Poaceae
Barren brome	*Bromus sterilis*	Poaceae
Basal stem rot, cucurbits	*Erwinia carotovora*	Eubacteriales: Enterobacteriaceae
Bean beetles	*Epilachna*	Coleoptera: Coccinellidae
Bean cutworm, Western	*Richia albicosta*	Lepidoptera: Noctuidae
Bean weevils	*Sitona*	Coleoptera: Curculionidae
Bearded oat	*Avena barbata*	Poaceae
Beet armyworm	*Spodoptera exigua*	Lepidoptera: Noctuidae
Beet armyworm	*Spodoptera litura*	Lepidoptera: Noctuidae
Beet cyst nematode	*Heterodera schachtii*	Nematoda: Heteroderidae
Beet leafminer	*Pegomya hyoscamni*	Diptera: Anthomyiidae
Beet leaf weevil	*Tanymecus pallidus*	Coleoptera: Curculionidae
Beetles	Coleoptera	Insecta
Begonia	*Begonia elatior*	Begoniaceae
Beneficial bacterium	*Agrobacterium radiobacter*	Eubacteriales: Rhizobiaceae
Bermuda grass	*Cynodon*	Poaceae
Bermuda grass	*Cynodon dactylon*	Poaceae
Bindweed, large	*Calystegia sepium*	Convolvulaceae
Bird-cherry aphid	*Rhopalosiphum padi*	Hemiptera: Aphididae
Bird skin mites	*Cnemidocoptes*	Acari: Sarcoptidae
Black-grass	*Alopecurus myosuroides*	Poaceae
Black bean aphid	*Aphis fabae*	Hemiptera: Aphididae
Black bent	*Agrostis gigantea*	Poaceae
Blackberries	*Rubus*	Rosaceae
Blackberry	*Rubus fruticosus*	Rosaceae
Blackberry rust	*Phragmidium violaceum*	Basidiomycetes: Uredinales
Black bindweed	*Fallopia convolvulus*	Polygonaceae
Blackcurrant gall mite	*Cecidophyopsis ribis*	Acari: Eriophyidae
Blackcurrant rust	*Cronartium ribicola*	Basidiomycetes: Uredinales
Black cutworm	*Agrotis ipsilon*	Lepidoptera: Noctuidae
Blackflies	*Simulium*	Diptera: Simuliidae
Black heart, apricot	*Verticillium dahliae*	Deuteromycetes: Moniliales
Black leaf streak, banana	*Mycosphaerella fijiensis*	Ascomycetes: Sphaeriales

English	Latin	Order and/or Family
Blackleg, beet crops	*Aphanomyces cochlioides*	Saprolegniales: Saprolegniacea
Blackleg, potato	*Erwinia carotvora*	Eubacteriales: Enterobacteriaceae
Black mould	*Cladosporium*	Deuteromycetes: Moniliales
Black nightshade	*Solanum nigrum*	Solanaceae
Black root rot, tobacco	*Thielaviopsis basicola*	Deuteromycetes: Moniliales
Black rot, apple	*Botryosphaeria obtusa* (= *Physalospora obtusa)*	Ascomycetes: Sphaeriales
Black rot, grapevine	*Guignardia bidwellii*	Ascomycetes: Sphaeriales
Black stem rust, grasses	*Puccinia graminis*	Basidiomycetes: Uredinales
Blackspot, roses	*Diplocarpon rosae*	Ascomycetes: Helotiales
Blight, capsicums	*Phytophthora capsici*	Oomycetes: Peronosporales
Blight, chestnut	*Cryphonectria parasitica*	Ascomycetes: Nectriaceae
Blight, damping off, foot rot, various hosts	*Phytophthora*	Oomycetes: Peronosporales
Blister blight, tea	*Exobasidium vexans*	Basidiomycetes: Exobasidiales
Blossom or pollen beetles	*Meligethes spp.*	Coleoptera: Nitidulidae
Blossom wilt, apple, plum	*Sclerotinia laxa*	Ascomycetes: Helotiales
Bluegrass webworm	*Parapediasia teterrella*	Lepidoptera: Crambidae
Blue mould, apple and pear	*Penicillium expansum*	Deuteromycetes: Moniliales
Blue mould, citrus	*Penicillium italicum*	Deuteromycetes: Moniliales
Blue mould, tobacco	*Peronospora tabacina* (= *Plasmopara tabacina)*	Oomycetes: Peronosporales
Blue rot, apple and pear	*Penicillium expansum*	Deuteromycetes: Moniliales
Bollworm, pink	*Pectinophora gossypiella*	Lepidoptera: Gelechiidae
Bollworm, spiny	*Earias*	Lepidoptera: Noctuidae
Bracken	*Pteridium aquilinum*	Filicales
Bracket chafer	*Phyllopertha horticola*	Coleoptera: Scarabaeidae
Bramble rust	*Phragmidium violaceum*	Basidiomycetes: Uredinales
Brambles	*Rubus*	Rosaceae
Bramble	*Rubus fructicosa*	Rosaceae
Branched bur-reed	*Sparganium erectum*	Sparganiaceae
Brassica cyst nematode	*Heterodera cruciferae*	Nematoda: Heteroderidae
Brassica gall and stem weevils	*Ceutorhynchus*	Coleoptera: Curculionidae
Brassica stem weevils	*Centorhynchus*	Coleoptera: Curculionidae
Broad mite	*Polyphagotarsonemus latus*	Acari: Tarsonemidae
Brooks spot, apple	*Mycosphaerella pomi*	Ascomycetes: Sphaeriales
Broomrape	*Orobanche* spp.	Orobanchaceae
Brown fleck of mushrooms	*Pseudomonas tolaasii*	Pseudomonadales: Pseudomonadaceae
Brown foot rot, cereals	*Gibberella* (= *Fusarium*)	Deuteromcetes: Hypocreales

English	Latin	Order and/or Family
Brown rot, apple, pear and plum	*Sclerotinia fructigena*	Ascomycetes: Helotiales
Brown rot, top fruit	*Sclerotinia fructicola*	Ascomycetes: Helotiales
Brown rot	*Pseudomonas solanacearum*	Pseudomonadales: Pseudomonadaceae
Brown rust, barley	*Puccinia hordei*	Basidiomycetes: Uredinales
Brown rust, chrysanthemum	*Puccinia chrysanthemi*	Basidiomycetes: Uredinales
Brown rust, wheat	*Puccinia triticina* (formerly *Puccinia recondite*)	Basidiomycetes: Uredinales
Brown soft scale	*Coccus hesperidum*	Hemiptera: Coccidae
Brown spot, peanut	*Mycosphaerella arachidis*	Ascomycetes: Sphaeriales
Brown spot, rice	*Cochliobolus miyabeanus*	Ascomycetes: Sphaeriales
Brown stripe, sugar cane	*Bipolaris stenospila*	Ascomycetes: Sphaeriales
Bt	*Bacillus thuringiensis*	Schizomycetes: Eubacteriales
Bugs	Heteroptera	Hemiptera
Bulb mites	*Rhizoglyphus callae, R. robini*	Acari: Acaridae
Bulb scale mite	*Steneotarsonemus laticeps*	Acari: Tarsonemidae
Bullrushes	*Typha*	Typhaceae
Bunt, stinking smut	*Tilletia caries*	Basidiomycetes: Ustilaginales
Burrowing nematode	*Radopholus similis*	Nematoda: Tylenchidae
Butt rot, conifers	*Fomes annosus*	Basidiomycetes: Agaricales
Buttercups	*Ranunculus*	Ranunculaceae
Butterflies	Lepidoptera	Insecta
Cabbage looper	*Trichoplusia ni*	Lepidoptera: Noctuidae
Cabbage moth	*Mamestra brassicae*	Lepidoptera: Noctuidae
Cabbage root fly	*Delia radicum*	Diptera: Anthomyiidae
Cabbage seed weevil	*Ceutorhynchus assimilis*	Coleoptera: Curculionidae
Cabbage stem weevil	*Ceutorhynchus quadridens*	Coleoptera: Curculionidae
Cabbage white butterflies	*Pieris*	Lepidoptera: Pieridae
Californian red scale	*Aonidiella aurantii*	Hemiptera: Diaspididae
Canadian pondweed	*Elodea canadensis*	Hydrocharitaceae
Canary grass, awned	*Phalaris paradoxa*	Poaceae
Canary grasses	*Phalaris*	Poaceae
Canegrub, greyback	*Dermolepida albohirtum*	Coleoptera: Scarabaeidae
Canker, apple and pear	*Diplodia pseudodiplodia*	Deuteromycetes: Sphaeropsidales
Canker, apple and pear	*Nectria galligena*	Hypocreales
Capsid bugs	Miridae	Heteroptera
Carmine spider mite	*Tetranychus cinnabarinus*	Acari: Tetranychidae
Carpenter ants	*Camponotus*	Hymenoptera: Formicidae
Carrot fly	*Psila rosae*	Diptera: Psilidae
Carrot leaf blight	*Alternaria dauci*	Deuteromycetes: Moniliales

References

English	Latin	Order and/or Family
Caterpillar predator	*Podisus maculiventris*	Heteroptera: Pentatomidae
Celery looper	*Syngrapha falcifera*	Lepidoptera: Noctuidae
Cercospora leaf spot, bean	*Cercospora zonata*	Deuteromycetes: Moniliales
Cereal flies	*Opomyza*	Diptera: Opomyzidae
Cereal leaf beetle	*Oulema melanopus*	Coleoptera: Chrysomelidae
Cereal looper	*Anagrapha falcifera*	Lepidoptera: Noctuidae
Charlock	*Sinapis arvensis*	Brassicaceae
Cherry tree borer	*Synanthedon hector*	Lepidoptera: Aegeriidae
Cherry leaf spot	*Blumeriella jaapii*	Ascomycetes: Heliotales
Chestnut blight	*Endothia parasitica*	Sphaeriales: Diaporthaceae
Chinch bug	*Blissus leucopterus*	Hemiptera: Lygaeidae
Chrysanthemum leaf miner	*Phytomyza syngenesiae*	Diptera: Agromyzidae
Chrysanthemum leaf miner parasitoid	*Dacnusa sibirica*	Hymenoptera: Braconidae
Chrysomelid beetles	Chrysomelidae	Coleoptera
Citrus aphid	*Aphis citricida*	Hemiptera: Aphididae
Citrus black scale	*Parlatoria ziziphi*	Hemiptera: Diaspididae
Citrus canker	*Xanthomonas citri*	Pseudomonadales: Pseudomonadaceae
Citrus flower moth	*Prays citri*	Lepidoptera: Yponomeutidae
Citrus mealybug	*Planococcus citri*	Hemiptera: Pseudococcidae
Citrus mussel scale	*Lepidosaphes (Cornuaspis) beckii*	Hemiptera: Coccidae
Citrus planthopper	*Metcalfa pruinesa*	Hemiptera: Fuloroidea
Citrus purple scale	*Lepidosaphes (Cornuaspis) beckii*	Hemiptera: Coccidae
Citrus red mite	*Panonychus citri*	Acari: Tetranychidae
Citrus red scale	*Chrysomphalus aonidum*	Hemiptera: Diaspididae
Citrus rust mite	*Phyllocoptruta oleivora*	Acari: Eriophyidae
Citrus weevil	*Pachnaeus litus*	Coleoptera: Curculionidae
Citrus whitefly	*Dialeurodes citri*	Hemiptera: Aleyrodidae
Cleavers	*Galium aparine*	Rubiaceae
Click beetles	Elateridae	Coleoptera
Clover bryobia mite	*Bryobia praetiosa*	Acari: Tetranychidae
Club-rushes	*Scirpus*	Cyperaceae
Clubroot, brassicas	*Plasmodiophora brassicae*	Oomycetes: Plasmodiophorales
Coccomycosis	*Blumeriella jaapii*	Ascomycetes: Helotiales
Cockchafer	*Melolontha hippocastani*	Coleoptera: Scarabaeidae
Cockchafer	*Melolontha melolontha*	Coleoptera: Scarabaeidae
Cockchafer, redheaded	*Adoryphorus couloni*	Coleoptera: Scarabaeidae
Cockchafers	*Melolontha*	Coleoptera: Scarabaeidae
Cocklebur	*Xanthium pennsylvanicum*	Asteraceae
Cocklebur, rough	*Xanthium strumarium*	Asteraceae
Cockroach, American	*Periplaneta americana*	Dictyoptera: Blattidae

English	Latin	Order and/or Family
Cockroach, German	*Blatella germanica*	Dictyoptera: Blattidae
Cockspur, rice	*Echinochloa oryzicola (= E. oryzoides)*	Poaceae
Cocoa capsid	*Distantiella theobroma*	Heteroptera: Miridae
Cocoa capsid	*Sahlbergella singularis*	Heteroptera: Miridae
Cocoa pod borer	*Conopomorpha cramerella*	Lepidoptera: Gracillariidae
Coconut mite	*Aceria guerreronis*	Acari: Eriophyidae
Coconut rhinoceros beetle	*Oryctes rhinoceros*	Coleoptera: Scarabaeidae
Codling moth	*Cydia pomonella*	Lepidoptera: Tortricidae
Coffee rust	*Hemileia vastatrix*	Basidiomycetes: Uredinales
Collar rot, apple	*Phytophthora cactorum*	Oomycetes: Peronosporales
Colorado beetle	*Leptinotarsa decemlineata*	Coleoptera: Chrysomelidae
Columbus grass	*Sorghum almum*	Poaceae
Common amaranth	*Amaranthus retroflexus*	Amaranthaceae
Common chickweed	*Stellaria media*	Caryophyllaceae
Common couch	*Elytrigia repens* (= *Elymus repens*)	Poaceae
Common green capsid	*Lygoceris pabulinus*	Heteroptera: Miridae
Common orache	*Atriplex patula*	Chenopodiaceae
Common scab of crops such as potato and beet	*Streptomyces scabies*	Schizomycetes: Actinomycetales
Cornborer, Mediterranean	*Sesamia nonagroides*	Lepidoptera: Pyralidae
Cornborer, pink	*Sesamia cretica*	Lepidoptera: Pyralidae
Corn marigold	*Chrysanthemum segetum*	Asteraceae
Corn rootworm	*Diabrotica undecimpunctata*	Coleoptera: Chrysomelidae
Corn rootworms	*Diabrotica*	Coleoptera: Chrysomelidae
Corn spurrey	*Spergula arvensis*	Caryophyllaceae
Cotton aphid	*Aphis gossypii*	Hemiptera: Aphididae
Cotton boll rot	*Gibberella fujikuroi*	Deuteromycetes: Hypocreales
Cotton boll weevil	*Anthonomus grandis*	Coleoptera: Curculionidae
Cotton leaf perforator	*Bucculatrix thurberiella*	Lepidoptera: Lyonetiidae
Cotton leaf worm	*Alabama argillacea*	Lepidoptera: Noctuidae
Cotton leafhoppers	*Empoasca*	Hemiptera: Cicadellidae
Cottony cushion scale	*Icerya purchasi*	Hemiptera: Margarodidae
Crabgrass	*Digitaria sanguinalis*	Poaceae
Crabgrass, tropical	*Digitaria adscendens (= D. ciliaris)*	Poaceae
Crabgrasses	*Digitaria*	Poaceae
Crane flies	*Tipula*	Diptera: Tipulidae
Cranesbills	*Geranium*	Geraniaceae
Creeping bent	*Agrostis stolonifera*	Poaceae
Creeping water primrose	*Ludwigia peploides*	Onagraceae
Crickets	*Cricetus*	Saltatoria: Gryllidae
Crickets, grasshoppers, etc.	Saltatoria	Insecta
Crown gall	*Agrobacterium tumefasciens*	Eubacteriales: Rhizobiaceae

English	Latin	Order and/or Family
Crown rot, apple	*Phytophthora cactorum*	Oomycetes: Peronsporales
Cucurbit anthracnose	*Colletotrichum lagenarium*	Ascomycetes: Melanconiales
Currant clearwing moth	*Synanthedon tipuliformis*	Lepidoptera: Aegeriidae
Cutworm	*Noctua pronuba*	Lepidoptera: Noctuidae
Cutworms	*Agrotis*	Lepidoptera: Noctuidae
Cutworms	*Euxoa*	Lepidoptera: Noctuidae
Cutworms	*Noctua*	Lepidoptera: Noctuidae
Cyst nematode, beet	*Heterodera schactii*	Nematoda: Heteroderidae
Cyst nematode, brassica	*Heterodera cruciferae*	Nematoda: Heteroderidae
Cyst nematode, pea	*Heterodera goettingiana*	Nematoda: Heteroderidae
Cyst nematode, soybean	*Heterodera glycines*	Nematoda: Heteroderidae
Cyst nematode, tobacco	*Globodera solanacearum*	Nematoda: Heteroderidae
Cyst nematodes	Heteroderidae	Nematoda
Damping off	*Rhizoctonia solani* (= *Thanetophorus cucumeris*)	Deuteromycetes: Agonomycetiales
Damping off, blight, foot rot, various hosts	*Phytophthora*	Oomycetes: Peronosporales
Damping off, rots etc., various hosts	*Pellicularia*	Basidiomycetes: Agaricales
Damping-off disease	*Thanetophorus cucumeris* (= *Rhizoctonia solani*)	Basidiomycetes: Stereales
Damson-hop aphid	*Phorodon humuli*	Hemiptera: Aphididae
Dark leaf spot, brassicas	*Alternaria brassicae*	Deuteromycetes: Moniliales
Dark leaf spot, brassicas	*Alternaria brassicicola*	Deuteromycetes: Moniliales
Dart moths	*Euxoa*	Lepidoptera: Noctuidae
Dayflower	*Commelina*	Commelinaceae
Dead arm, grapevine	*Phomopsis viticola*	Deuteromycetes: Sphaeropsidales
Diamondback moth	*Plutella xylostella*	Lepidoptera: Yponomeutidae
Diffuse knapweed	*Centaurea diffusa*	Asteraceae
Docks and sorrels	*Rumex*	Polygonaceae
Dodder	*Cuscuta*	Convolvulaceae
Dodder, Australian	*Cuscuta australis*	Convolvulaceae
Dodder, Chinese	*Cuscuta chinensis*	Convolvulaceae
Dodder, large	*Cuscuta europaea*	Convolvulaceae
Dollar spot, turf	*Sclerotinia homeocarpa*	Ascomycetes: Helotiales
Douglas fir	*Pseudotsuga menziesii*	Pinaceae
Douglas fir beetle	*Dendroctonus pseudotsuga*	Coleoptera: Scolytidae
Douglas fir beetle	*Ips typographus*	Coleoptera: Ipidae
Douglas fir tussock moth	*Orgyia pseudotsugata*	Lepidoptera: Lymantriidae
Downy mildew, brassicae	*Peronospora parasitica*	Oomycetes: Peronosporales
Downy mildew, cereals	*Scerophthora macrospora*	Oomycetes: Peronosporales
Downy mildew, cucurbits	*Pseudoperonospora cubensis*	Oomycetes: Peronosporales

English	Latin	Order and/or Family
Downy mildew, grapevine	*Plasmopara viticola*	Oomycetes: Peronosporales
Downy mildew, hop	*Pseudoperonospora humuli*	Oomycetes: Peronosporales
Downy mildew, lettuce	*Bremia lactucae*	Oomycetes: Peronosporales
Downy mildew, sorghum	*Peronosclerospora*	Oomycetes: Peronosporales
Downy mildew, wheat	*Scerophthora*	Oomycetes: Peronosporales
Downy mildews	*Peronospora*	Oomycetes: Peronosporales
Downy mildews, e.g. on pearl millet	*Sclerospora*	Oomycetes: Peronosporales
Downy mildews, various hosts	*Plasmopara*	Oomycetes: Peronosporales
Dry bubble, mushrooms	*Verticillium fungicola*	Deuteromycetes: Moniliales
Dry rot, postharvest rot	*Fusarium coeruleum*	Deuteromycetes: Moniliales
Dutch elm disease	*Ceratocystis ulmi*	Ascomycetes: Sphaeriales
Ear blight, cereals	*Gibberella* (= *Fusarium*)	Deuteromycetes: Hypocreales
Egyptian cotton leafworm	*Spodoptera littoralis*	Lepidoptera: Noctuidae
Elodea, Florida	*Hydrilla verticillata*	Hydrocharitaceae
Eriophyid mites	Eriophyidae	Acari
Eucalyptus psyllid	*Ctenarytaina eucalypti*	Hemiptera: Psyllidae
Eucalyptus snout beetle	*Gonipterus gibberus*	Coleoptera: Curculionidae
European cornborer	*Ostrinia nubilalis*	Lepidoptera: Pyralidae
European grapevine moth	*Lobesia botrana*	Lepidoptera: Tortricidae
European pine sawfly	*Neodiprion sertifer*	Hymenoptera: Diprionidae
European pine shoot moth	*Rhyacionia buoliana*	Lepidoptera: Tortricidae
Eye-spot, cereals	*Pseudocercosporella herpotrichoides*	Deuteromycetes: Moniliales
Fairy rings	*Marasmius oreades* and other species	Agaricales
Fall armyworm	*Spodoptera frugiperda*	Lepidoptera: Noctuidae
Fall panicum	*Panicum dichotomiflorum*	Poaceae
False oat-grass	*Arrhenatherum elatius*	Poaceae
Fat hen	*Chenopodium album*	Chenopodiaceae
Field bindweed	*Convolvulus arvensis*	Convolvulaceae
Field pansy	*Viola arvensis*	Violaceae
Field slug	*Decoceras*	Mollusca: Gastropoda
Fire ants	*Solenopsis*	Hymenoptera: Formicidae
Fire blight of pome fruit	*Erwinia amylovora*	Eubacteriales: Enterobacteriaceae
Flea beetle	*Phyllotreta striolata*	Coleoptera: Chrysomelidae
Flea beetles	*Chaetocnema*	Coleoptera: Chrysomelidae
Flea beetles	*Phyllotreta*	Coleoptera: Chrysomelidae
Flies	Diptera	Insecta
Flour beetles	Cucijidae	Coleoptera

English	Latin	Order and/or Family
Fly speck disease, apple	*Schizothyrium pomi*	Ascomycetes: Hemisphaeriales
Foot rot, blight, damping off, various hosts	*Phytophthora*	Oomycetes: Peronosporales
Foot rot, cereals and grasses	*Cochliobolus sativus*	Ascomycetes: Sphaeriales
Foot rot, various hosts	*Aphanomyces*	Oomycetes: Saprolegniales
Foot rot, various hosts	*Rhizoctonia*	Deuteromycetes: Agonomycetiales
Formosan termite	*Coptotermes formosanus*	Isoptera: Rhinotermitidae
Four-leaved water clover	*Marsilea*	Marsileaceae
Foxtail, giant	*Setaria faberi*	Poaceae
Foxtail grasses	*Setaria*	Poaceae
Foxtail, green	*Setaria viridis*	Poaceae
Foxtail, yellow	*Setaria glauca* (= *S. lutescens*)	Poaceae
Fringe rushes	*Fimbristylis*	Cyperaceae
Frit fly	*Oscinella frit*	Diptera: Chloropidae
Frog eye, soybean	*Cercosporidium* (*C. sojinum* = *Cercospora sojina*)	Deuteromycetes: Moniliales
Fruit flies	Drosophilidae	Diptera
Fruit flies	*Dacus*	Diptera: Tephritidae
Fruit flies	*Drosophila*	Diptera: Drosophilidae
Fruit moth, plum	*Cydia funebrana*	Lepidoptera: Tortricidae
Fruit rot, strawberry	*Mucor*	Mucorales
Fruit rot, various hosts	*Botrytis cinerea*	Deuteromycetes: Moniliales
Fruit tree red spider mite	*Panonychus ulmi*	Acari: Tetranychidae
Fruit tree red spider mite predator	*Amblyseius degenerans*	Mesostigmata: Phytoseiidae
Fruit tree red spider mite predator	*Typhlodromus pyri*	Mesostigmata: Phytoseiidae
Fuchsia	*Fuchsia hybrida*	Onagraceae
Fungal vector of rhizomania virus	*Polymyxa betae*	Oomycetes: Plasmodiophorales
Fungus gnat predator	*Hypoaspis aculeifer*	Mesostigmata: Laelapidae
Fungus gnats	Sciaridae	Diptera
Fusarium foot and root rots, various hosts	*Fusarium culmorum*	Deuteromycetes: Moniliales
Fusarium wilt, various hosts	*Fusarium oxysporum*	Deuteromycetes: Moniliales
Gall midges	Cecidomyiidae	Diptera
Gangrene, potato	*Phoma exigua* var. *foventa*	Deuteromycetes: Sphaeropsidales
Garden chafer	*Phyllopertha horticola*	Coleoptera: Scarabaeidae
Garden wireworms	*Athous*	Coleoptera: Elateridae
German cockroach	*Blatella germanica*	Dictyoptera: Blattidae
Giant knotweed	*Reynoutria sachalinensis* (= *Polygonum sachalinense*)	Polygonaceae

English	Latin	Order and/or Family
Glasshouse potato aphid	*Aulacorthum solani*	Hemiptera: Aphididae
Glasshouse whitefly	*Trialeurodes vaporariorum*	Hemiptera: Aleyrodidae
Glasshouse whitefly parasitoid	*Encarsia formosa*	Hymenoptera: Aphelinidae
Gloeosporium rot, apple	*Gloeosporium*	Deuteromycetes: Sphaeriales
Gloeosporium rot, apple	*Glomerella cingulata*	Deuteromycetes: Sphaeriales
Glume blotch, wheat	*Leptosphaeria nodorum* (= *Septoria nodorum)*	Ascomycetes: Sphaeriales
Golden chalcid	*Aphytis melinus*	Hymenoptera: Aphelinidae
Gorse	*Ulex europaeus*	Fabaceae
Gooseberry bryobia mite	*Bryobia ribis*	Acari: Tetranychidae
Goosegrass	*Eleusine indica*	Poaceae
Grain aphid	*Sitobium avenae*	Hemiptera: Aphididae
Grain beetles	*Cryptolestes*	Coleoptera: Cucujidae
Grape berry moth	*Eupoecilia ambiguella*	Lepidoptera: Cochylidae
Grape leafhopper	*Erythroneura elegentala*	Hemiptera: Cicadellidae
Grass and cereal flies	*Opomyza*	Diptera: Opomyzidae
Grasshoppers	Acarididae	Saltatoria
Grasshoppers, crickets, etc.	Saltatoria	Insecta
Grass moth	*Chrysoteuchia caliginosellus* (= *Crambus caliginosellus)*	Lepidoptera: Pyralidae
Greasy blotch, carnation	*Zygophiala jamaicensis*	Deuteromycetes: Moniliales
Green leafhopper	*Empoasca fabae*	Hemiptera: Cicadellidae
Green leafhoppers	*Nephotettix*	Hemiptera: Cicadellidae
Green mould, citrus	*Penicillium digitatum*	Deuteromycetes: Moniliales
Green rice leafhopper	*Nephotettix impicticepts*	Hemiptera: Cicadellidae
Green rice leafhopper	*Nephotettix cincticeps*	Hemiptera: Cicadellidae
Green rice leafhopper, tropical	*Nephotettix aigropictus*	Hemiptera: Cicadellidae
Gypsy moth	*Lymantria dispar*	Lepidoptera: Lymantriidae
Halo blight, bean	*Pseudomonas phaseolicola*	Pseudomonadales: Pseudomonadaceae
Hay bacillus	*Bacillus subtilis*	Schizomycetes: Eubacteriales
Head smut, maize	*Sphacelotheca reiliana*	Basidiomycetes: Ustilaginales
Helmet scale	*Saissetia coffeae*	Hemiptera: Coccidae
Helminthosporium blight, rice	*Helminthosporium oryzae*	Deuteromycetes: Moniliales
Hemispherical scale	*Saissetia coffeae*	Hemiptera: Coccidae
Hemp sesbania	*Sesbania exaltata*	Fabaceae
Hickory shuckworm	*Cydia caryana*	Lepidoptera: Tortricidae
Honey fungus	*Armillaria mellea*	Basidiomycetes: Agaricales
Hoppers	Hemiptera	Hemiptera

English	Latin	Order and/or Family
Hornweed, common	*Ceratophyllum demersum*	Ceratophyllaceae
Housefly	*Musca domestica*	Diptera: Muscidae
House mosquito	*Culex fatigans* (= *C. quinquefasciatus*)	Diptera: Culicidae
Hyperparasite of fungi	*Ampelomyces quisqualis*	Deuteromycetes: Sphaeropsidales
Indian meal moth	*Plodia interpunctella*	Lepidoptera: Pyralidae
Ivy	*Hedera helix*	Araliaceae
Japanese beetle	*Popillia japonicus*	Coleoptera: Scarabaeidae
Japanese bullrush	*Scirpus juncoides*	Cyperaceae
Japanese knotweed	*Reynoutria japonica* (= *Polygonum cuspidatum*)	Polygonaceae
Jimson weed	*Datura stramonium*	Solanaceae
Johnson grass	*Sorghum halepense*	Poaceae
Joint vetches	*Aeschynomene*	Fabaceae
Kanzawa spider mite	*Tetranychus kanzawi*	Acari: Tetranychidae
Knapweed, diffuse	*Centaurea diffusa*	Asteraceae
Knapweed, spotted	*Centaurea maculosa*	Asteraceae
Knapweeds	*Centaurea*	Asteraceae
Knot grass	*Polygonum aviculare*	Polygonaceae
Knotweeds	*Polygonum*	Polygonaceae
Koa seed worm	*Crytophlebia illepida*	Lepidoptera: Tortricidae
Kyllinga, green	*Cyperus brevifolius*	Cyperaceae
Ladybird	*Harmonia axyridis*	Coleoptera: Coccinellidae
Ladybird	*Hippodamia convergens*	Coleoptera: Coccinellidae
Ladybug	*Harmonia axyridis*	Coleoptera: Coccinellidae
Ladybug	*Hippodamia convergens*	Coleoptera: Coccinellidae
Large fruit flies	Tephritidae	Diptera
Large white butterfly	*Pieris brassicae*	Lepidoptera: Pieridae
Late blight, potato, tomato	*Phytophthora infestans*	Oomycetes: Peronosporales
Late-flowering cyperus	*Cyperus serotinus*	Cyperaceae
Leaf and glume spots, various hosts	*Septoria*	Sphaeropsidales
Leaf and pod spot, pea	*Ascochyta pinodes*	Deuteromycetes: Sphaeropsidales
Leaf and pod spot, pea	*Ascochyta pisi*	Deuteromycetes: Sphaeropsidales
Leaf blight, rice	*Xanthomonas oryzae*	Pseudomonadales: Pseudomonadaceae
Leaf blister moth, pear	*Leucoptera malifoliella*	Lepidoptera: Lyonetiidae
Leaf blotch, barley and rye	*Rhynchosporium secalis*	Deuteromycetes: Moniliales

English	Latin	Order and/or Family
Leaf blotches, etc., various hosts	*Marssonina*	Melanconiales
Leaf miners	*Agromyza*	Diptera: Agromyzidae
Leaf miners	*Liriomyza*	Diptera: Agromyzidae
Leaf miners	*Phytomyza*	Diptera: Agromyzidae
Leaf-mining moths	*Leucoptera*	Lepidoptera: Lyonetiidae
Leaf-mining moths	*Phyllonorycter*	Lepidoptera: Gracillariidae
Leaf mould, tomato	*Fulvia fulva*	Deuteromycetes: Hyphales
Leaf moulds	*Fulvia*	Deuteromycetes: Hyphales
Leaf roller	*Archips podanus*	Lepidoptera: Tortricidae
Leaf roller	*Pandemis heparana*	Lepidoptera: Totricidae
Leaf roller	*Platynota stultana*	Lepidoptera: Totricidae
Leaf rollers	*Adoxophyes*	Lepidoptera: Torticidae
Leaf rollers	*Humona*	Lepidoptera: Torticidae
Leaf scorch, apple	*Gymnosporangium*	Basidiomycetes: Uredinales
Leaf scorch, strawberry	*Diplocarpon earliana*	Ascomycetes: Helotiales
Leaf smuts, various hosts	*Urocystis*	Basidiomycetes: Ustilaginales
Leaf spot	*Alternaria alternata*	Deuteromycetes: Moniliales
Leaf spot	*Curvularia*	Deuteromycetes: Moniliales
Leaf spot, apple	*Botryosphaeria obtusa* (= *Physalospora obtusa)*	Ascomycetes: Sphaeriales
Leaf spot, bean	*Ascochyta fabae*	Deuteromycetes: Sphaeropsidales
Leaf spot, beet crops	*Cercospora beticola*	Deuteromycetes: Moniliales
Leaf spot, beet crops	*Ramularia beticola*	Deuteromycetes: Moniliales
Leaf spot, currant and gooseberry	*Pseudopeziza ribis*	Ascomycetes: Helotiales
Leaf spot diseases, various hosts	*Mycosphaerella*	Ascomycetes: Sphaeriales
Leaf spot, melon	*Corynespora melonis*	Deuteromycetes: Moniliales
Leaf spot, soybean	*Cercosporidium* (= *C. sojinum* = *Cercospora sojina*)	Deuteromycetes: Moniliales
Leaf spot, sunflower	*Diaporthe helianthi*	Ascomycetes: Sphaeriales
Leaf spots, grasses	*Rhynchosporium*	Deuteromycetes: Moniliales
Leaf spots and bacterial blights, various hosts	*Pseudomonas*	Pseudomonadales: Pseudomonadaceae
Leaf spots, various	*Alternaria*	Deuteromycetes: Moniliales
Leaf spots, various	*Cercospora*	Deuteromycetes: Moniliales
Leaf spots, various	*Ramularia*	Deuteromycetes: Moniliales
Leaf spots, various hosts	*Ascochyta*	Deuteromycetes: Sphaeropsidales
Leaf stripe, barley	*Pyrenophora graminea*	Ascomycetes: Sphaeriales
Leafhopper	*Empoasca decipiens*	Hemiptera: Cicadellidae
Leafhopper	*Hauptidia maraccana*	Hemiptera: Cicadellidae
Leafhopper egg parasitoid	*Anagrus atomus*	Hymenoptera: Mymaridae

English	Latin	Order and/or Family
Leafworm, Egyptian cotton	*Spodoptera littoralis*	Lepidoptera: Noctuidae
Leafy spurge	*Euphorbia esula*	Euphorbiaceae
Leatherjackets	*Tipula*	Diptera: Tipulidae
Lemon-shaped cyst nematodes	*Heterodera*	Nematoda: Heteroderidae
Leopard moth	*Zeuzera pyrina*	Lepidoptera: Cossidae
Lepidopteran egg parasites	*Trichogramma*	Hymenoptera: Trichogrammatidae
Lepidopteran parasitic wasp	*Cotesia*	Hymenoptera: Aphidiidae
Lesser armyworm	*Spodoptera exigua*	Lepidoptera: Noctuidae
Lesser grain borer	*Rhyzopertha dominica*	Coleoptera: Bostrichidae
Lesser peach tree borer	*Synanthedon pictipes*	Lepidoptera: Aegeriidae
Light leaf spot, brassicas	*Pyrenopeziza brassicae*	Ascomycetes: Helotiales
Liverworts	Bryophyta	Bryophyta
Locusts	Acrididae	Saltatoria
Long-tailed mealybug	*Pseudococcus longispinus*	Hemiptera: Pseudococcidae
Looper, soybean	*Pseudoplusia includens*	Lepidoptera: Noctuidae
Loose silky-bent	*Apera spica-venti*	Poaceae
Loose smut, barley and wheat	*Ustilago nuda*	Basidiomycetes: Ustilaginales
Macadamia nut borer	*Cryptophlebia ombrodelta*	Lepidoptera: Tortricidae
Macrotermites	*Macrotermes*	Uniramia: Macrotermitinae
Macrotermites	*Microtermes*	Uniramia: Macrotermitinae
Macrotermites	*Odontotermes*	Uniramia: Macrotermitinae
Maggots	*Delia* and *Hylemya*	Diptera: Anthomyiidae
Maggots	*Hylemya*	Diptera: Anthomyiidae
Maize stalk borer	*Diatraea saccharalis*	Lepidoptera: Pyralidae
Maize stalk rot	*Gibberella fujikuroi*	Deuteromycetes: Hypocreales
Maize weevil	*Sitophilus zeamais*	Coleoptera: Curculionidae
Mallow	*Malva*	Malvaceae
Mange mites	*Chorioptes*	Acari: Psoroptidae
Mangold flea beetle	*Chaetocnema concinna*	Coleoptera: Chrysomelidae
Mangold fly	*Pegomya hyoscamni*	Diptera: Anthomyiidae
Mangrove	*Rhizophora mangle*	Rhizophoraceae
Marigold, African	*Tagetes erecta*	Asteraceae
Mayweed, scentless	*Tripleurospermum maritimum* (= *Matricaria inodora*)	Asteraceae
Mayweeds (some)	*Chamomilla*	Asteraceae
Mayweeds (some)	*Matricaria*	Asteraceae
McDaniel's spider mite	*Tetranychus mcdanieli*	Acari: Tetranychidae
Meadowgrass, annual	*Poa annua*	Poaceae
Meadowgrass, rough	*Poa trivialis*	Poaceae
Meadowgrasses	*Poa*	Poaceae

English	Latin	Order and/or Family
Mealworms	*Alphitobius*	Coleoptera: Tenebrionidae
Mealybug, citrus	*Planococcus citri*	Hemiptera: Pseudococcidae
Mealybug, long-tailed	*Pseudococcus longispinus*	Hemiptera: Pseudococcidae
Mealybug, obscure	*Pseudococcus affinis*	Hemiptera: Pseudococcidae
Mealybug parasite	*Leptomastix dactylopii*	Hymenoptera: Encrytidae
Mealybug predator	*Cryptolaemus montrouzieri*	Coleoptera: Coccinelidae
Mealybugs	*Pseudococcus*	Hemiptera: Pseudococcidae
Mediterranean black scale	*Saissetia oleae*	Hemiptera: Coccidae
Mediterranean fruit fly	*Ceratitis capitata*	Diptera: Tephritidae
Melanosis, citrus	*Diaporthe citri*	Ascomycetes: Sphaeriales
Melon and cotton aphid	*Aphis gossypii*	Hemiptera: Aphididae
Melon fly	*Dacus cucurbitae*	Diptera: Tephritidae
Mexican bean beetle	*Epilachna varivestis*	Coleoptera: Coccinellidae
Milkweed vine	*Morrenia odorata*	Asclepiadaceae
Millet, Texas	*Panicum texanum*	Poaceae
Millipedes	Diplopoda	Myriapoda
Minute black ladybird	*Stethorus punctillum*	Coleoptera: Coccinellidae
Minute pirate bugs	*Orius*	Hemiptera: Anthocoridae
Mite predator	*Feltiella acarisuga*	Diptera: Cecidomyiidae
Mite predator	*Galendromus occidentalis*	Acarina: Phytoseiidae
Mite, avocado brown	*Oligonychus punicae*	Acari: Tetranychidae
Mite, bulb scale	*Steneotarsenemus laticeps*	Acari: Tetranychidae
Mite, carmine spider	*Tetranychus cinnabarinus*	Acari: Tetranychidae
Mite, Kanzawa spider	*Tetranychus kanzawi*	Acari: Tetranychidae
Mite, McDaniel's spider	*Tetranychus mcdanieli*	Acari: Tetranychidae
Mite, Pacific spider	*Tetranychus pacificus*	Acari: Tetranychidae
Mite, persea	*Oligonychus perseae*	Acari: Tetranychidae
Mite, six-spotted	*Eotetranychus sexmaculatus*	Acari: Tetranychidae
Mite, southern red	*Oligonychus illicis*	Acari: Tetranychidae
Mite, spruce spider	*Oligonychus ununguis*	Acari: Tetranychidae
Mite, storage	*Acarus siro*	Acarina: Acaridae
Mite, storage	*Lepidoglyphus (Glycyphagus) destructor*	Acarina: Acaridae
Mite, strawberry	*Tetranychus atlanticus*	Acari: Tetranychidae
Mite, two-spotted spider	*Tetranychus urticae*	Acari: Tetranychidae
Mite, willamette	*Eotetranychus willamettei*	Acari: Tetranychidae
Mites	*Aculops*	Acari: Eriophyidae
Mites	*Calepitrimerus*	Acari: Eriophyidae
Mites	*Eriophyes*	Acari: Eriophyidae
Mites	*Phyllocoptes*	Acari: Eriophyidae
Mites	*Vasates*	Acari: Eriophyidae
Mites, spider	Tetranychidae	Acari
Mites, spider	*Tetranychus* spp.	Acari: Tetranychidae

English	Latin	Order and/or Family
Mock cypress	*Kochia scoparia*	Chenopodiaceae
Mole cricket	*Gryllotalpa gryllotalpa*	Orthoptera: Gryllotalpidae
Mole cricket	*Scapteriscus vicinus*	Orthoptera: Gryllotalpidae
Mole crickets	*Gryllotalpa*	Orthoptera: Gryllotalpidae
Monilinia leaf blight, apple	*Monilinia mali*	Deuteromycetes: Moniliales
Morning glory, ivyleaf	*Ipomoea hederacea*	Convolvulaceae
Morning glory, tall	*Ipomoea purpurea*	Convolvulaceae
Mosquitos	*Aedes*, *Anopheles*, *Coquillettidea*, *Culex*, *Culiseta*, *Deinocerites*, *Eretmapodites*, *Haemagogus*, *Mansonia*, *Opifex*, *Orthopodomyia*, *Psorophora*, *Sabethes*, *Uranotaenia* and *Wyeomyia* spp.	Diptera: Culicidae
Moss (typical)	*Ceratodon purpureus*	Bryophyta
Moths	Lepidoptera	Insecta
Mountain pine beetle	*Dendroctonus ponderosae*	Coleoptera: Scolytidae
Mugwort	*Artemisia vulgaris*	Asteraceae
Mushroom cecid	*Heteropeza pygmaea*	Diptera: Cecidomyiidae
Mushroom sciarid	*Lycoriella auripila*	Diptera: Sciaridae
Navel orangeworm	*Amylois transitella*	Lepidoptera: Pyralidae
Neck rot, lettuce	*Sclerotinia minor*	Ascomycetes: Helotiales
Neck rot, onion	*Botrytis allii*	Deuteromycetes: Moniliales
Needle nematodes	*Longidorus*	Nematoda
Neem tree	*Azadirachta indica*	Meliaceae
Net blotch, barley	*Pyrenophora teres*	Ascomycetes: Sphaeriales
Nettle, common	*Urtica dioica*	Urticaceae
Nettle, small	*Urtica urens*	Urticaceae
Nettles	*Urtica*	Urticaceae
New Zealand grass grub	*Costelytra zealandrica*	Coleoptera: Scarabaeidae
N-fixing bacteria	*Nitrosomonas*	Bacteria
Nipplewort	*Lapsana communis*	Asteraceae
Noctuid moths	Noctuidae	Lepidoptera
Northern joint vetch	*Aeschynomene virginica*	Fabaceae
Northern leaf blight, maize	*Helminthosporium turcicum*	Deuteromycetes: Moniliales
Nutgrass	*Cyperus rotundus*	Cyperaceae
Nutsedges	*Cyperus*	Cyperaceae
Oat, sterile	*Avena sterilis*	Poaceae
Oats (wild and cultivated)	*Avena*	Poaceae
Old World bollworm	*Helicoverpa armigera*	Lepidoptera: Noctuidae
Olive fruit fly	*Bactrocera oleae* (= *Dacus oleae*)	Diptera: Tephritidae

English	Latin	Order and/or Family
Olive moth	*Prays oleae*	Lepidoptera: Yponomeutidae
Olive scale, black	*Saissetia oleae*	Hemiptera: Coccidae
Onion couch	*Arrhenatherum elatius*	Poaceae
Onion thrips	*Thrips tabaci*	Thysanoptera: Thripidae
Orangeworm, navel	*Amylois transitella*	Lepidoptera: Pyralidae
Oriental fruit moth	*Grapholitha molesta*	Lepidoptera: Tortricidae
Oriental fruit tree moth	*Carposina niponensis*	Lepidoptera: Tortricidae
Oriental red scale	*Aonidiella orientalis*	Hemiptera: Diaspididae
Oriental tobacco budworm	*Helicoverpa assulta*	Lepidoptera: Noctuidae
Pacific spider mite	*Tetranychus pacificus*	Acari: Tetranychidae
Pale persicaria	*Polygonum lapathifolium*	Polygonaceae
Palm beetle, American	*Rhynchopherus palmarum*	Coleoptera: Curculionidae
Palm scale	*Chrysomphalus dictyospermi*	Hemiptera: Diaspididae
Palm weevil, red	*Rhynchophorus ferrugineus*	Coleoptera: Curculionidae
Panic grasses	*Panicum*	Poaceae
Para grass	*Brachiaria mutica*	Poaceae
Parasitic yeasts	*Candida*	Ascomycetes: Endomycetales
Pea aphid	*Acyrthosiphon pisum*	Hemiptera: Aphididae
Pea cyst nematode	*Heterodera goettingiana*	Nematoda: Heteroderidae
Pea moth	*Cydia nigricana*	Lepidoptera: Tortricidae
Pea weevils	*Sitona*	Coleoptera: Curculionidae
Peach leaf-curl	*Taphrina deformans*	Ascomycetes: Taphrinales
Peach tree borer	*Anarsia lineatella*	Lepidoptera: Gelechiidae
Peach tree borer	*Sanninoidea exitiosa*	Lepidoptera: Aegeriidae
Peach-potato aphid	*Myzus persicae*	Hemiptera: Aphididae
Pear leaf blister moth	*Leucoptera malifoliella*	Lepidoptera: Lyonetiidae
Pear psylla	*Cacopsylla pyri*	Hemiptera: Psyllidae
Pear psylla	*Cacopsylla pyricola*	Hemiptera: Psyllidae
Pear rust	*Gymnosporangium fuscum*	Basidiomycetes: Uredinales
Pear rust mite	*Epitrimerus pyri*	Acari: Eriophyidae
Pearly green lacewing	*Chrysoperla carnea*	Neuroptera: Chrysopidae
Penicillium rots	*Penicillium*	Deuteromycetes: Moniliales
Persicaria	*Polygonum persicaria*	Polygonaceae
Petunia	*Petunia*	Solanaceae
Pharaoh's ant	*Monomorium pharaonis*	Hymenoptera: Formicidae
Pickerel weed	*Monochoria vaginalis*	Pontederiaceae
Pimpernel, false	*Lindernia procumbens*	Scrophulariaceae
Pine false webworms	*Acantholyda erythrocephala*	Hymenoptera: Pamphiliidae
Pine processionary caterpillar	*Thaumetopoea pityocampa*	Lepidoptera: Thaumetopoeidae

English	Latin	Order and/or Family
Pine sawfly	*Neodiprion lecontei*	Hymenoptera: Diprionidae
Pink bollworm	*Pectinophora gossypiella*	Lepidoptera: Gelechiidae
Pitch canker disease	*Fusarium moniliforme* var. *subglutinans*	Deuteromycetes: Moniliales
Plantains	*Plantago*	Plantaginaceae
Planthopper, citrus	*Metcalfa pruinosa*	Hemiptera: Fulgoroidea
Planthopper, small brown	*Laodelphax striatella*	Hemiptera: Delphacidae
Planthopper, white-banded	*Sogatella furcifera*	Hemiptera: Delphacidae
Planthoppers	*Nilaparvata*	Hemiptera: Delphacidae
Plum rust	*Tranzchelia pruni-spinosae*	Basidiomycetes: Uredinales
Pod rot, cocoa	*Monilia roreri*	Deuteromycetes: Moniliales
Pollen beetle	*Meligethes aeneus*	Coleoptera: Nitidulidae
Pollen beetles	*Meligethes*	Coleoptera: Nitidulidae
Pondweed, American	*Potamogeton distinctus*	Potamogetonaceae
Pondweeds	*Potamogeton*	Potamogetonaceae
Poplars	*Populus*	Salicaceae
Poppies	*Papaver*	Papaveraceae
Postharvest rots	*Fusarium coeruleum*	Deuteromycetes: Moniliales
Postharvest rots	*Rhizopus*	Phycomycetes: Mucorales
Postharvest rots, various hosts	*Sclerotium*	Deuteromycetes: Agonomycetales
Potato aphid	*Macrosiphum euphorbiae*	Hemiptera: Aphididae
Potato cyst nematodes	*Globodera*	Nematoda: Heteroderidae
Potato leafhopper	*Eupterycyba jucunda*	Hemiptera: Cicadellidae
Potato moth	*Phthorimaea operculella*	Lepidoptera: Gelechiidae
Powdery mildew, apple	*Podosphaera leucotricha*	Ascomycetes: Erysiphales
Powdery mildew, beet crops	*Erysiphe betae*	Ascomycetes: Erysiphales
Powdery mildew, cereals and grasses	*Erysiphe graminis*	Ascomycetes: Erysiphales
Powdery mildew, cucurbits	*Erysiphe cichoracearum*	Ascomycetes: Erysiphales
Powdery mildew, cucurbits	*Sphaerotheca fuliginea*	Ascomycetes: Erysiphales
Powdery mildew, grapevine	*Uncinula necator*	Ascomycetes: Erysiphales
Powdery mildew, peppers	*Leveillula*	Ascomycetes: Erysiphales
Powdery mildew, rose	*Sphaerotheca pannosa*	Ascomycetes: Erysiphales
Powdery mildew, tomato	*Leveillula taurica*	Ascomycetes: Erysiphales
Powdery mildew, tomato	*Oidiopsis taurica*	Deuteromycetes: Moniliales
Powdery mildew, various hosts	*Erysiphe*	Ascomycetes: Erysiphales
Powdery mildew, various hosts	*Phyllactinia*	Ascomycetes: Erysiphales
Powdery mildew, various hosts	*Podosphaera*	Ascomycetes: Erysiphales
Powdery mildew, various hosts	*Sphaerotheca*	Ascomycetes: Erysiphales
Predacious midges	Cecidomyiidae	Diptera
Predatory mite	*Amblyseius barkeri*	Mesostigmata: Phytoseiidae
Predatory mite	*Amblyseius californicus*	Mesostigmata: Phytoseiidae
Predatory mite	*Amblyseius cucumeris*	Mesostigmata: Phytoseiidae

English	Latin	Order and/or Family
Predatory mite	*Typhlodromus occidentalis*	Mesostigmata: Phytoseiidae
Prickly pear cactus	*Opuntia lindheimerii*	Cactaceae
Prickly sida	*Sida spinosa*	Malvaceae
Psyllid, eucalyptus	*Ctenarytaina eucalypti*	Hemiptera: Psyllidae
Psyllids	*Psylla*	Hemiptera: Psyllidae
Puncture vine	*Tribulus terrestris*	Zygophyllaceae
Purslane	*Portulaca oleracea*	Portulacaceae
Purslanes	*Portulaca*	Portulacaceae
Pygmy mangold beetle	*Atomaria linearis*	Coleoptera: Cryptophagidae
Pyrethrum daisy	*Tanacetum cinerariaefolium*	Compositeae
Quackgrass	*Elytrigia repens* (= *Elymus repens*)	Poaceae
Ragweed, common	*Ambrosia artemisifolia*	Asteraceae
Ragwort, common	*Senecio jacobaea*	Asteraceae
Ragwort, tansy	*Senecio jacobaea*	Asteraceae
Rape	*Brassica napus*	Brassicaceae
Ray blight, chrysanthemum	*Didymella ligulicola*	Ascomycetes: Sphaeriales
Red core, strawberry	*Phytophthora fragariae*	Phycomycetes: Peronosporales
Red crevice tea mite	*Brevipalpus phoenicis*	Acari: Tenuipalpidae
Red dead-nettle	*Lamium purpureum*	Labiatae
Red scale	*Aspidiotus nerii*	Hemiptera: Diaspididae
Red scale parasite	*Aphytis lignanensis*	Hymenoptera: Aphelinidae
Red spider mites	*Panonychus*	Acari: Tetranychidae
Red thread	*Laetisaria fuciformis*	Basidiomycetes: Aphyllophorales
Redheaded cockchafer	*Adoryphorus couloni*	Coleoptera: Scarabaeidae
Redroot pigweed	*Amaranthus retroflexus*	Amaranthaceae
Redshank	*Polygonum persicaria*	Polygonaceae
Reniform nematode	*Rotylenchulus reniformis*	Nematoda
Rhizomania virus, fungal vector	*Polymyxa betae*	Oomycetes: Plasmodiophorales
Rhododendron	*Rhododendron ponticum*	Ericaceae
Rice blast (imperfect stage)	*Pyricularia oryzae*	Deuteromycetes: Moniliales
Rice blast (perfect stage)	*Magnaporthe grisea*	Ascomycetes: Magnaporthales
Rice brown planthopper	*Nilaparvata lugens*	Hemiptera: Delphacidae
Rice leaf beetle	*Oulema oryzae*	Coleoptera: Chrysomelidae
Rice leaf roller	*Cnaphalocrocis medinalis*	Lepidoptera: Pyralidae
Rice leaf scald	*Monographella nivalis*	Ascomycetes: Sphaeriales
Rice sheath blight	*Pellicularia sasakii*	Basidiomycetes: Agaricales
Rice stalk borer	*Chilo suppressalis*	Lepidoptera: Pyralidae

English	Latin	Order and/or Family
Rice stem borer	*Chilo plejadellus*	Lepidoptera: Pyralidae
Rice stem borer	*Chilo suppressalis*	Lepidoptera: Pyralidae
Rice stem rot	*Magnaporthe salvinii* (see also *Sclerotium oryzae* and *Nakataea sigmoidae*)	Ascomycetes: Pezizales
Rice water weevil	*Lissorhoptrus oryzophilus*	Coleoptera: Curculionidae
Rice weevil	*Sitophilus oryzae*	Coleoptera: Curculionidae
Rice weevil	*Sitophilus zeamais*	Coleoptera: Curculionidae
Ring-spot, brassicas	*Mycosphaerella brassicicola*	Ascomycetes: Sphaeriales
Root and stem rots, various	*Phoma*	Deuteromycetes: Sphaeriales
Root flies	*Delia* (= *Hylemya*)	Diptera: Anthomyiidae
Root-knot nematodes	*Meloidogyne*	Nematoda
Root-lesion nematodes	*Pratylenchus*	Nematoda: Hoplolaimidae
Root rot, brassicas	*Phytophthora megasperma*	Oomycetes: Peronosporales
Root rot, cereals and grasses	*Cochliobolus sativus*	Ascomycetes: Sphaeriales
Root rot, tomato	*Colletotrichum coccodes*	Ascomycetes: Melanconiales
Root rot, various hosts	*Aphanomyces*	Oomycetes: Saprolegniales
Root rot, various hosts	*Rhizoctonia*	Deutermycetes: Agonomycetiales
Root rots, various	*Pythium*	Oomycetes: Peronosporales
Root rots, various	*Rhizoctonia solani* (= *Thanetophorus cucumeris*)	Deutermycetes: Agonomycetiales
Rose aphid	*Macrosiphum rosae*	Hemiptera: Aphididae
Rose thrips	*Thrips fuscipennis*	Thysanoptera: Thripidae
Rot, various crops	*Phytophthora palmivora*	Oomycetes: Peronosporales
Rots of stems, storage organs, etc., various crops	*Sclerotinia sclerotiorum*	Ascomycetes: Helotiales
Rots, damping off, etc. various hosts	*Pellicularia*	Basidiomycetes: Agaricales
Rots, ear blights and wilts, various hosts (imperfect fungi with perfect stages in various genera)	*Fusarium*	Deuteromycetes: Moniliales
Rots, various	*Monilia*	Deuteromycetes: Moniliales
Rots, various hosts	*Sclerotium rolfsii*	Deuteromycetes: Agonomycetales
Rough cocklebur	*Xanthium strumarium*	Asteraceae
Roughseed bullrush	*Scirpus mucronatus*	Cyperaceae
Rubbery rot, potato	*Geotrichum candidum*	Deuteromycetes: Moniliales
Runch	*Raphanus raphanistrum*	Brassicaceae
Rush, flowering	*Butomus umbellatus*	Butomaceae
Rush skeletonweed	*Chondrilla juncea*	Asteraceae
Russian thistle	*Salsola kali*	Chenopodiaceae
Rust fungi	Uredinales	Basidiomycetes
Rust mite, apple	*Aculus schlechtendali*	Acari: Eriophyidae

English	Latin	Order and/or Family
Rust mite, citrus	*Phyllocaptrata oleivora*	Acari: Eriophyidae
Rust mites	*Aculus*	Acari: Eriophyidae
Rust mites	*Phyllocoptruta*	Acari: Eriophyidae
Rust, beet crops	*Uromyces betae*	Basidiomycetes: Uredinales
Rust, blackberry	*Phragmidium violaceum*	Basidiomycetes: Uredinales
Rust, roses	*Phragmidium mucronatum*	Basidiomycetes: Uredinales
Rust, soybean	*Phakopsora pachyrhizi*	Basidiomycetes: Uredinales
Rust, various hosts	*Puccinia*	Basidiomycetes: Uredinales
Rust, various hosts	*Gymnosporangium*	Basidiomycetes: Uredinales
Rusts, various crops	*Uromyces*	Basidiomycetes: Uredinales
Ryegrass, annual	*Lolium rigidum*	Poaceae
Ryegrass, Italian	*Lolium multiflorum*	Poaceae
Ryegrass, perennial	*Lolium perenne*	Poaceae
Ryegrass, Wimmera	*Lolium rigidum*	Poaceae
Ryegrasses	*Lolium*	Poaceae
St John's wort	*Hypericum perforatum*	Guttifereae
St John's wort, pale	*Hypericum montanum*	Guttifereae
San José scale	*Quadraspidiotus perniciosus*	Hemiptera: Coccidae
Saprophytic fungi	*Paecilomyces*	Deuteromycetes: Hyphales
Saramatta grass	*Ischaemum rugosum*	Poaceae
Sawflies	*Diprion*	Hymenoptera: Diprionidae
Scab, apple	*Venturia inaequalis*	Ascomycetes: Sphaeriales
Scab, cereals	*Gibberella zeae*	Deuteromycetes: Hypocreales
Scab, cereals; brown foot rot and ear blight and other cereal diseases	*Gibberella* (= *Fusarium*)	Deuteromycetes: Hypocreales
Scab, citrus	*Elsinoe fawcettii*	Ascomycetes: Myriangiales
Scab, pear	*Venturia pirina*	Ascomycetes: Sphaeriales
Scale insect	*Didesmococcus brevipes*	Hemiptera: Coccidae
Scale insects	Coccidae	Hemiptera
Scale insects	*Coccus* spp.	Hemiptera: Coccidae
Scale insects	Diaspididae	Hemiptera
Scale insects	Margarodidae	Hemiptera
Scale insect parasite	*Metaphycus bartletti*	Hymenoptera: Encrytidae
Scale insect parasite	*Metaphycus helvolus*	Hymenoptera: Encrytidae
Scale, black olive	*Saissetia oleae*	Hemiptera: Coccidae
Scale, citrus black	*Parlatoria ziziphi*	Hemiptera: Diaspididae
Scale, cottony cushion	*Icerya purchasi*	Hemiptera: Margarodidae
Scale, *Euonymus*	*Unaspis euonymi*	Hemiptera: Coccidae
Scale, *Euonymus alatus*	*Lepidosaphes yangicola*	Hemiptera: Coccidae
Scale, helmet	*Saissetia coffeae*	Hemiptera: Coccidae
Scale, hemispherical	*Saissetia coffeae*	Hemiptera: Coccidae
Scale, Mediterranean black	*Saissetia oleae*	Hemiptera: Coccidae

English	Latin	Order and/or Family
Scale, siebold	*Lepidosaphes yangicola*	Hemiptera: Coccidae
Scentless mayweed	*Tripleurospermum maritimum* (= *Matricaria inodora)*	Asteraceae
Sciarid flies	*Bradysia*	Diptera: Sciaridae
Sciarid flies	*Lycoriella*	Diptera: Sciaridae
Sciarid flies	*Sciara*	Diptera: Sciaridae
Sciarid flies	Sciaridae	Diptera
Sciarid fly predator	*Hypoaspis miles*	Mesostigmata: Laelapidae
Sclerotinia rots, various hosts	*Sclerotinia*	Ascomycetes: Helotiales
Scotchbroom	*Sarothamnus scoparius*	Fabaceae
Scuttle flies	*Megaselia*	Diptera: Phoridae
Scuttle flies	Phoridae	Diptera
Sea club-rush	*Scirpus maritimus*	Cyperaceae
Sea-rush	*Juncus maritimus*	Juncaceae
Sedges	*Carex*	Cyperaceae
Seed-eating ants	*Monomorium*	Hymenoptera: Formicidae
Septoria leaf spot, wheat	*Mycosphaerella graminicola*	Ascomycetes: Sphaeriales
Sharp eyespot, cereals	*Ceratobasidium cereale*	Basidiomycetes: Tulasnellales
Shattercane	*Sorghum bicolor*	Poaceae
Shepherd's purse	*Capsella bursa-pastoris*	Brassicaceae
Shoot borer, Eastern pine	*Eucosma gloriola*	Lepidoptera: Tortricidae
Shothole, prunus	*Stigmina carpophila*	Deuteromycetes: Hyphales
Siam weed	*Eupatorium odoratum* (= *Chromolaena odorata*)	Asteraceae
Sickle pod	*Cassia obtusifolia*	Fabaceae
Sigatoka	*Mycosphaerella musicola*	Ascomycetes: Sphaeriales
Silver leaf	*Chondrostereum purpureum*	Agaricales: Agaricaceae
Silver scurf, potato	*Helminthosporium solani*	Deuteromycetes: Moniliales
Silvery moth	*Plusia*	Lepidoptera: Noctuidae
Six-spined ips	*Ips sexdentatus*	Coleoptera: Ipidae
Six-toothed spruce bark beetle	*Pityogenes chalcographus*	Coleoptera: Scolytidae
Skin spot, potato	*Polyscytalum pustulans*	Deuteromycetes: Hyphales
Slugs	Gastropoda	Mollusca
Small brown planthopper	*Laodelphax striatella*	Hemiptera: Delphacidae
Small white butterfly	*Pieris rapae*	Lepidoptera: Pieridae
Smaller European elm bark beetle	*Scolytus multistratus*	Coleoptera: Scolytidae
Smartweed	*Polygonum persicaria*	Polygonaceae
Smooth sowthistle	*Sonchus oleraceus*	Asteraceae
Smooth witchgrass	*Panicum dichotomiflorum*	Poaceae
Smut diseases, various hosts	*Ustilago*	Basidiomycetes: Ustilaginales
Smut, various hosts	*Tilletia*	Basidiomycetes: Ustilaginales

English	Latin	Order and/or Family
Snails	Gastropoda	Mollusca
Snow mould, grasses and cereals	*Microdochium nivalis*	Deuteromycetes: Moniliales
Snow rot, cereals	*Typhula incarnata*	Basidiomycetes: Agaricales
Sooty blotch, apple, pear and citrus	*Gloeodes pomigena*	Deuteromycetes: Sphaeropsidales
Sooty mould	*Cladosporium*	Deuteromycetes: Moniliales
Sorghum grasses	*Sorghum*	Poaceae
Sorrels	*Rumex*	Polygonaceae
South American leaf miner	*Liriomyza huidobrensis*	Diptera: Agromyzidae
Southern pine beetle	*Dendroctonus frontalis*	Coleoptera: Scolytidae
Southern root-knot nematode	*Meloidogyne incognita*	Nematoda
Southwestern cornborer	*Diatraea grandiosella*	Lepidoptera: Pyralidae
Soybean cyst nematode	*Heterodera glycines*	Nematoda: Heteroderidae
Soybean looper	*Anticarsia gemmatalis*	Lepidoptera: Noctuidae
Soybean looper	*Pseudoplusia includens*	Lepidoptera: Noctuidae
Speedwell, common field	*Veronica persica*	Scrophulariaceae
Speedwell, ivy-leaved	*Veronica hederifolia*	Scrophulariaceae
Speedwell, slender	*Veronica filiformis*	Scrophulariaceae
Speedwells	*Veronica*	Scrophulariaceae
Spider mite predator	*Amblyseius fallacis*	Mesostigmata: Phytoseiidae
Spider mites	*Tetranychus*	Acari: Tetranychidae
Spike rush	*Eleocharis acicularis*	Cyperaceae
Spiny bollworms	*Earias*	Lepidoptera: Noctuidae
Spiny sida	*Sida spinosa*	Malvaceae
Spiral nematodes	*Helicotylenchus*	Nematoda: Tylenchidae
Spotted knapweed	*Centaurea maculosa*	Asteraceae
Spotted spurge	*Euphorbia maculata*	Euphorbiaceae
Sprangletop grasses	*Leptochloa*	Poaceae
Sprangletop, bearded	*Leptochloa fascicularis* (= *Diplachne fascicularis)*	Poaceae
Sprangletop, red	*Leptochloa chinensis*	Poaceae
Spruce beetle	*Dendroctonus rufipennis*	Coleoptera: Scolytidae
Spur blight, cane fruit	*Didymella applanata*	Ascomycetes: Sphaeriales
Spurge, leafy	*Euphorbia esula*	Euphorbiaceae
Spurge, spotted	*Euphorbia maculate*	Euphorbiaceae
Stalk rots, various hosts	*Diplodia*	Deuteromycetes: Sphaeropsidales
Stem rot, rice	*Magnaporthe salvinii* (see also *Sclerotium oryzae* and *Nakataea sigmoidae*)	Ascomycetes: Pezizales
Storage mite	*Acarus siro*	Acarina: Acaridae
Storage mite	*Lepidoglyphus* (*Glycyphagus*) *destructor*	Acarina: Acaridae
Star grasses	*Cynodon*	Poaceae

References

English	Latin	Order and/or Family
Star-thistle, purple	*Centaurea calcitrapa*	Asteraceae
Star-thistle, yellow	*Centaurea solstitialis*	Asteraceae
Stem borers	*Chilo*	Lepidoptera: Pyralidae
Stem canker, sunflowers	*Diaporthe helianthi*	Ascomycetes: Sphaeriales
Stem canker fungi, various hosts	*Diaporthe*	Ascomycetes: Sphaeriales
Stem nematode	*Ditylenchus dipsaci*	Nematoda: Tylenchidae
Sting nematode	*Belonolaimus longicausatus*	Nematoda
Stinking smut	*Tilletia caries*	Basidiomycetes: Ustilaginales
Storage fungi	*Aspergillus*	Deuteromycetes: Moniliales
Strangler vine	*Morrenia odorata*	Asclepiadaceae
Stubby-root nematodes	*Trichodorus*	Nematoda
Sugar beet root maggot	*Tetanops myopaeformis*	Diptera: Otitidae
Sugar cane borer	*Diatreae saccharalis*	Lepidoptera: Pyralidae
Sugar cane rootstalk borer	*Diaprepes abbreviatus*	Coleoptera: Curculionidae
Sugarcane spittle bug	*Mahanarva postica*	Hemiptera: Cercopidae
Sumac	*Rhus aromatica*	Anacardiacea
Summer fruit tortrix moth	*Adoxophyes orana*	Lepidoptera: Tortricidae
Sunflower	*Helianthus annuus*	Asteraceae
Symphilids	*Symphyla*	Myriapoda
Tan spot, wheat	*Pyrenophora tritici-vulgaris*	Ascomycetes: Sphaeriales
Tarnished plant bug	*Lygus pabulinus*	Heteroptera: Miridae
Tarnished plant bug, Southern	*Lygus lineolaris*	Heteroptera: Miridae
Tarnished plant bug, Western	*Lygus herperus*	Heteroptera: Miridae
Tarsonemid mites	*Tarsonemus* (= *Phytonemus*, in part)	Acari: Tarsonemidae
Tea leaf roller	Caloptilia theivora	Lepidoptera: Gracillariidae
Tea tortrix	*Homona magnanima*	Lepidoptera: Tortricidae
Teaberry	*Gaultheria procumbens*	Ericacea
Termites	*Coptotermes*	Isopteran: Rhinotermitidae
Tetranychid mites	*Eutetranychus*	Acari: Tetranychidae
Texas citrus mite	*Eutetranychus banksi*	Acari: Tetranychidae
Thistle, creeping	*Cirsium arvense*	Asteraceae
Thistle, Russian	*Salsola kali*	Chenopodiaceae
Thistles	*Carduus*	Asteraceae
Thorn apple	*Datura stramonium*	Solanaceae
Thrips	*Thrips*	Thysanoptera: Thripidae
Thrips	*Thrips palmi*	Thysanoptera: Thripidae
Thrips, avocado	*Scirtothrips persea*	Thysanoptera: Thripidae
Thrips, flower	*Frankliniella intonsa*	Thysanoptera: Thripidae
Thrips, legume flower	*Megalurothrips sjostedti*	Thysanoptera: Thripidae
Thrips, New Zealand flower	*Thrips obscuratus*	Thysanoptera: Thripidae

English	Latin	Order and/or Family
Thrips, onion	*Thrips tabaci*	Thysanoptera: Thripidae
Thrips, palm	*Parthenothrips dracaenae*	Thysanoptera: Thripidae
Thrips, rose	*Thrips fuscipennis*	Thysanoptera: Thripidae
Thrips, Western flower	*Frankliniella occidentalis*	Thysanoptera: Thripidae
Thyme	*Thymus vulgaris*	Labiatae
Tobacco	*Nicotiana rustica*	Solanaceae
Tobacco budworm	*Heliothis virescens*	Lepidoptera: Noctuidae
Tobacco cyst nematode	*Globodera solanacearum*	Nematoda: Heteroderidae
Tobacco flea beetle	*Epitrix hirtipennis*	Coleoptera: Chrysomelidae
Tobacco whitefly	*Bemisia tabaci*	Hemiptera: Aleyrodidae
Tomato canker	*Clavibacter michiganensis*	Eubacteriales
Tomato leaf miner	*Liriomyza bryoniae*	Diptera: Agromyzidae
Tomato looper	*Chrysodeixis chalcites*	Lepidoptera: Noctuidae
Tomato pinworm	*Keiferia lycopersicella*	Lepidoptera: Gelechiidae
Tomato powdery mildew	*Leveillula taurica*	Ascomycetes: Erysiphales
Tomato powdery mildew	*Oidiopsis taurica*	Deuteromycetes: Moniliales
Torpedo grass	*Panicum repens*	Poaceae
Tortrix moths	*Tortrix*	Lepidoptera: Tortricidae
Tortrix moths	*Adoxophyes*	Lepidoptera: Tortricidae
Tortrix moths	*Homona*	Lepidoptera: Tortricidae
Tropical green rice leafhopper	*Nephotettix nigropictus*	Hemiptera: Cicadellidae
True crickets	Gryllidae	Orthoptera: Gryllotalpidae
Tufted apple moth	*Platynota idaeusalis*	Lepidoptera: Tortricidae
Turnip gall weevil	*Ceutorhynchus pleurostigmata*	Coleoptera: Curculionidae
Turnip moth	*Agrotis segetum*	Lepidoptera: Noctuidae
Tussock moth, Douglas fir	*Orgyia pseudotsugata*	Lepidoptera: Lymantriidae
Tussock moth, whitemarked	*Orgyia leucostigma*	Lepidoptera: Lymantriidae
Two-spotted spider mite	*Tetranychus urticae*	Acari: Tetranychidae
Two-spotted spider mite predator	*Phytoseiulus persimilis*	Mesostigmata: Phytoseiidae
Umbrella plant	*Cyperus difformis*	Cyperaceae
Valsa canker of apple	*Valsa ceratosperma*	Ascomycetes: Sphaeriales
Velvet bean caterpillar	*Anticarsia gemmatalis*	Lepidoptera: Noctuidae
Velvetleaf	*Abutilon theophrasti*	Malvaceae
Verticillium wilt, various hosts	*Verticillium*	Deuteromycetes: Moniliales
Vine weevil	*Otiorhynchus sulcatus*	Coleoptera: Curculionidae
Wandering Jew	*Commelina*	Commelinaceae
Warehouse moth	*Ephestia elutella*	Lepidoptera: Pyralidae
Water clover, four-leaved	*Marsilea*	Marsileaceae
Water duckweed	*Pistia stratiotes*	Araceae
Water hyacinth	*Eichhornia crassipes*	Pontederiaceae

English	Latin	Order and/or Family
Water primrose, creeping	*Ludwigia peploides*	Onagraceae
Water primroses	*Jussiaea*	Onagraceae
Water purslane	*Ludwigia peploides*	Onagraceae
Water weed	*Elodea canadensis*	Hydrocharitaceae
Weevils	Curculionidae	Coleoptera
Welsh chafer	*Hoplia philanthus*	Coleoptera: Scarabaeidae
Western balsam bark beetle	*Dryocoetes confusus*	Coleoptera: Scolytidae
Western bean cutworm	*Richia albicosta*	Lepidoptera: Noctuidae
Western flower thrips	*Frankliniella occidentalis*	Thysanoptera: Thripidae
Wheat bulb fly	*Delia coarctata*	Diptera: Anthomyiidae
White-backed planthopper	*Sogatella furcifera*	Hemiptera: Delphacidae
White blister	*Albugo candida*	Oomycetes: Peronosporales
Whitefly	*Bemisia*	Hemiptera: Aleyrodidae
Whitefly	*Aleurothrixus floccosus*	Hemiptera: Aleyrodidae
Whitefly parasite	*Eretmocerus* sp. nr. *californicus* (= *Eretmocerus eremicus*)	Hymenoptera: Aphelinidae
Whitefly predatory beetle	Delphasus pusillus	Coleoptera: Coccinellidae
Whitefly, citrus	*Dialeurodes citri*	Hemiptera: Aleyrodidae
Whitefly, glasshouse	*Trialeurodes vaporariorum*	Hemiptera: Aleyrodidae
Whitefly, silverleaf	*Bemisia argentifolii*	Hemiptera: Aleyrodidae
Whitefly, tobacco	*Bemisia tabaci*	Hemiptera: Aleyrodidae
White grubs	*Hoplochelis marginalis*	Coleoptera: Scarabaeidae
White leaf spot, oilseed rape	*Pseudocercosporella capsellae*	Deuteromycetes: Hyphales
White leaf spot, strawberry	*Mycosphaerella fragariae*	Ascomycetes: Sphaeriales
White mould, mushrooms	*Mycogone perniciosa*	Deuteromycetes: Moniliales
White muscardine	*Beauvaria bassiana*	Deuteromycetes: Moniliales
White mustard	*Sinapis alba*	Brassicaceae
White rot, onion	*Sclerotium cepivorum*	Deuteromycetes: Agonomycetales
White rot, timber	*Ganoderma*	Basidiomycetes: Agaricales
Wild fire of tobacco and soybean	*Pseudomonas tabaci*	Pseudomonadales: Pseudomonadaceae
Wild oat	*Avena fatua*	Poaceae
Wild oat, winter	*Avena ludoviciana*	Poaceae
Wild pansies	*Viola*	Violaceae
Wild radish	*Raphanus raphanistrum*	Brassicaceae
Wilt	*Verticillium albo-atrum*	Deuteromycetes: Moniliales
Wimmera ryegrass	*Lolium rigidum*	Poaceae
Winterberry	*Gaultheria procumbens*	Ericacea
Wireworms	*Agriotes*	Coleoptera: Elateridae
Witches' broom	*Crinipellis perniciosa*	Basidiomycetes: Agaricales
Woolly aphid	*Eriosoma lanigerum*	Hemiptera: Pemphigidae
Wormwood	*Artemisia vulgaris*	Compositeae

English	Latin	Order and/or Family
Yeasts	Endomycetales	Ascomycetes
Yellow birch	*Betula lutea*	Betulaceae
Yellow cereal fly	*Opomyza florum*	Diptera: Opomyzidae
Yellow fever mosquito	*Aedes aegypti*	Diptera: Culicidae
Yellow-headed spruce sawfly	*Pikonema alaskensis*	Hymenoptera: Tenthredinidae
Yellow nutsedge	*Cyperus esculentus*	Cyperaceae
Yellow rust, cereals	*Puccinia striiformis*	Basidiomycetes: Uredinales
Yellow underwing moth	*Noctua pronuba*	Lepidoptera: Noctuidae
Yew	*Taxus baccata*	Taxaceae

Directory of companies

Parts of names given in bold represent the short form of the company name which is used in the text of Main Entries.

AgBio
9915 Raleigh Street,
Westminster,
CO 80030,
USA
Tel: 1 303 469 9221
Fax: 1 303 469 9598

AgBioChem Inc.
3 Fleetwood Court,
Orinda,
CA 94563,
USA
Tel: 1 510 254 0789
Fax: 1 925 254 0186

AgraQuest Inc.
1105 Kennedy Place, No. 4,
Davis,
CA 95616-1272,
USA
Tel: 1 530 750 0150
Fax: 1 530 750 0153
e-mail: agraquest@agraquest.com

AgraTech Seeds Inc.
5559 North 500 West,
McCordsville,
IN 46055-9998,
USA
Tel: 1 317 335 333
Fax: 1 317 335 9260

AgResearch
Level 1A,
6 Viaduct Harbour Avenue,
Maritime Square,
Private Bag 92080,
Auckland,
NEW ZEALAND
Tel: 64 9 968 9100
Fax: 64 9 968 9101
e-mail: agresearch@agresearch.co.nz
Home page: www.agresearch.co.nz

AgrEvo
See **Bayer CropScience**

Agriculture and Agri-Food Canada
1341 Baseline Road,
Ottawa,
Ontario K1A 0C5,
CANADA
Tel: 1 613 773 1000
Fax: 1 613 773 2772
e-mail: info@agr.gc.ca
Home page: www.agr.gc.ca

Agricultural Research Initiatives Inc.
700 Research Center Blvd,
Fayetteville,
AR 72701,
USA
e-mail: dkellyc@alltel.net

References

Agriculture Sciences Inc.
3601 Garden Brook,
Dallas,
TX 75234,
USA
Tel: 1 972 243 8930
Fax: 1 972 406 1125

Agriculture Solutions
PO Box 141,
Strong,
ME 04983,
USA
Tel: 207-684-3939
e-mail: info@agriculturesolutions.com

Agripro Seeds Inc.
PO Box 2962,
6700 Antioch,
Shawnee Mission,
KS 66201-1362,
USA
Tel: 1 913 384 4940
Fax: 1 913 384 0208

Agritope
See **Exelixis** Inc.

Agro-Kanesho Co., Ltd
Akasaka Shasta-East 7th Fl.,
4-2-19 Akasaka,
Minato-Ku,
Tokyo, 107-0052,
JAPAN
Tel: 81 3 5570 4711
Fax: 81 3 5570 4708
Home page: www.agrokanesho.co.jp/

AGSCI
See **Agriculture Sciences** Inc.

Agventure Seeds Inc.
207 North Seventh Street,
PO Box 29,
Kentland,
IN 47951,
USA
Tel: 1 219 474 5557
e-mail: agventure@agventure.com

American **Cyanamid** Co.
See **BASF**

Aragonesas Agro, S.A.
Po Recoletos 27,
28004 Madrid,
SPAIN
Tel: 34 91 5853800
Fax: 34 91 5852310

Asgrow Seed Co.
3000 Westown Parkway,
West Des Moines,
IA 50266,
USA
Tel: 1 515 224 4200
Fax: 1 515 224 4262

Aventis
See **Bayer CropScience**

BASF AG
Crop Protection Division,
Agricultural Center,
Postfach 120,
D-67114 Limburgerhof,
GERMANY
Tel: 49 621 60 0
Fax: 49 621 60 27144
Home page: www.basf.com
Agricultural products: www.basf.de/en/produkte/gesundheit/pflanzen

Bayer Corporation
Agriculture Div.,
8400 Hawthorn Rd,
PO Box 4913,
Kansas City,
MO 64120-0013,
USA
Tel: 1 816 242 2000
Fax: 1 816 242 2738

Bayer CropScience
Business Group Crop Protection,
Development/Regulatory Affairs,
Agrochemical Center Monheim,
D-51368 Leverkusen,
GERMANY
Tel: 49 2173 38 3280
Fax: 49 2173 38 3564
Home page: www.bayer.com/
Agricultural products: www.bayer.com/en/tk/cropscience.php

Becks Superior **Hybrids** Inc.
6767 East 276 Street,
Atlanta,
IN 47096,
USA
Tel: 1 317 984 2325
Fax: 1 317 984 3508

Bio Huma Netics Inc.
201 South Roosevelt,
Chandler,
AZ 85226,
USA
Tel: 1 480 961 1220
Fax: 1 480 961 3501
e-mail: info@biohumanetics.com
Home page: www.biohumanetics.com

Biosystemes France
Parc d'Activités des Bellevues,
BP 227,
95614 Cergy-Pontoise CEDEX,
FRANCE
Tel: 33 34 48 99 26
Fax: 33 34 48 99 27

Biotech International
PO Box 1539,
9109 Main Street,
Needville,
TX 77461,
USA
Tel: 1 979 793 78 80
Fax: 1 979 793 49 11
e-mail: info@biotechintl.com
Home page: www.biotechintl.com

Callahan Seeds
1122 East 169th Street,
Westfield,
IN 46074,
USA
Tel: 1 317 896 5551
Fax: 1 317 896 9209

Campbell Seeds Inc.
1375 North 800W,
Tipton,
IN 46072,
USA
Fax: 1 765 963 2047

Cargill Hybrid Seeds
PO Box 5645,
Minneapolis,
MN 55440-2399,
USA
Tel: 1 612 337 9100
Fax: 1 612 742 7233

Chemgro Seeds Co.
PO Box 218,
East Petersburg,
PA 17520,
USA
Tel: 1 717 569 3296
Fax: 1 717 560 0117

Coated Seed Ltd
See **Wrightson Seeds** Ltd

Countrymark Cooperative Inc.
PO Box 2500,
Bloomington,
IL 61702-2500,
USA
Tel: 1 309 557 6399
Fax: 1 309 557 6860

Croplan Genetics
PO Box 64089,
MS 690 St Paul,
MN 55164-0089,
USA
Tel: 1 612 451 5458

Cyanamid
See **BASF** AG

Dairyland Seed Co., Inc.
PO Box 958,
West Bend,
WI 53095-0958,
USA
Fax: 1 414 626 2281

DeKalb Genetics Corp.
3100 Sycamore Road,
Dekalb,
IL 60115,
USA
Tel: 1 815 758 9273
Fax: 1 815 756 2672

Deltapine Seed
Box 157,
Scott,
MS 38732,
USA
Tel: 1 601 742 4000
Fax: 1 601 742 4055

Dow AgroScience
9330 Zionsville Rd,
Indianapolis,
IN 46268-1054,
USA
Tel: 1 317 337 4974
Fax: 1 317 337 7344
e-mail: info@dowagro.com
Home page: www.dowagro.com/

DowElanco
See **Dow AgroScience**

DNA Plant Technology Corporation
6701 San Pablo Avenue,
Oakland,
CA 94608,
USA
Tel: 1 510 547 2395
Fax: 1 510 450 9395
(No longer trading)

Du Pont
See E. I. **DuPont** de Nemours

E. I. **DuPont** de Nemours
Du Pont Agricultural Products,
Walker's Mill,
Barley Mill Plaza,
Wilmington,
DE 19880,
USA
Tel: 1 800 441 7515
Fax: 1 302 992 6470
Home page: www.dupont.com/index.html
Agricultural products: www.dupont.com/ag/

Exelixis Inc.
210 East Grand Avenue,
PO Box 511,
So. San Francisco,
CA 94080-0511,
USA
Tel: 1 650 837 7000
Fax: 1 650 837 8300

Florigene Pty Ltd
1 Park Drive,
Bundoora,
Victoria - 3083,
AUSTRALIA
Tel: 61 3 9243 3800
Fax: 61 3 9243 3888
e-mail: florigene@florigene.com.au

Garst Seed Co.
PO Box 300,
Coon Rapids,
IA 50058,
USA
Tel: 1 712 684 3243
Fax: 1 712 684 3300

GlycoGenesys, Inc.
Park Square Building 31,
St. James Avenue, 8th Floor,
Boston,
MA 02116,
USA
Tel: 1 617 422 0674
Fax: 1 617 422 0675
Home page: www.glycogenesys.com/

Golden Harvest Seeds Inc.
220 Eldorado Road,
Suite E,
Bloomington,
IL 61704-3544,
USA
Tel: 1 309 664 0558
Fax: 1 309 664 0984

Growmark Inc.
PO Box 2500,
Bloomington,
IL 61702-2500,
USA
Tel: 1 309 557 6399
Fax: 1 309 557 6860

Gustafson Inc.
1400 Preston Road,
Suite 400,
Plano,
TX 75093,
USA
Tel: 1 972 985 8877
Fax: 1 972 985 1696

Gutwein Seeds
RR 1 Box 40,
Francesville,
IN 47945,
USA
Tel: 1 219 567 9141
Fax: 1 219 567 2645

Hoechst Schering **AgrEvo** GmbH
See **Bayer CropScience**

Hoegemeyer Hybrids
1755 Hoegemeyer Road,
Hooper,
NE 68031,
USA
Tel: 1 402 654 3399
Fax: 1 402 654 3342

Hoffman Seeds Inc.
144 Main Street,
Landisville,
PA 17538-1297,
USA
Tel: 1 717 898 2261
Fax: 1 717 898 9458

Igene Biotechnology Inc.
9110 Red Branch Road,
Columbia,
MD 21045-2024,
USA
Tel: 1 410 997 2599
Fax: 1 410 730 0540
e-mail: igene@igene.com

Interstate Seed Co.
PO Box 338,
West Fargo,
ND 58078,
USA
Tel: 1 701 282 7338
Fax: 1 701 282 8218

Latham Seeds
131 180th Street,
Alexander,
IA 50420,
USA
Tel: 1 515 692 3258
Fax: 1 515 692 3250

Merck & Co., Inc.
See **Syngenta**

Merschman Seeds
103 Avenue D,
West Point,
IA 52656,
USA
Tel: 1 319 837 6111
Fax: 1 319 837 6104

Midwest Seed Genetics
PO Box 518,
Carroll,
IA 51401,
USA
Tel: 1 712 792 6691
Fax: 1 712 792 6725

Monsanto Co.
Crop Protection,
800 N. Lindbergh Blvd,
St. Louis,
MO 63167,
USA
Tel: 1 314 694 1000
Fax: 1 314 694 7625
Home page: www.monsanto.com

Monsanto Europe S.A.
Avenue de Tervuren 270-272,
B-1150 Brussels,
BELGIUM

Mycogen Corp.
See **Dow AgroScience**

Mycogen Crop Protection
See **Dow AgroScience**

Mycogen
1340 Corporate Center Curve,
St Paul,
MN 55121-1428,
USA
Tel: 1 612 405 5954
Fax: 1 612 405 5957

NatureMark Potatoes
300 East Mallard Drive,
Suite 220,
Boise,
ID 83706,
USA
Tel: 1 208 389 2236
Fax: 1 208 309 2280

NC+ Hybrids
PO Box 4408,
Lincoln,
NE 68504,
USA
Tel: 1 402 467 2517
Fax: 1 402 467 4217

Nematech
Plant Nematology Lab.,
Centre for Plant Sciences, LIBA,
Irene Manton Building,
University of Leeds,
Leeds, LS2 9JT,
UK
e-mail: pbiu@leeds.ac.uk

NOR-AM
See **Bayer CropScience**

Novartis Crop Protection AG
See **Syngenta**

Novartis Seeds Inc.
7500 Olson Memorial Highway,
Golden Valley,
MN 55427,
USA
Tel: 1 612 593 7189
Fax: 1 612 593 7203

Papaya Administrative Committee
Hilo,
HI
USA
Tel: 1 808 969 1160
e-mail: papayas@aloha.net

Patriot Seeds Inc.
208 South Worrell,
PO Box 97,
Bowen,
IL 62316-0097,
USA
Fax: 1 217 842 5209

Paymaster Cottonseed
1301 East 50th Street,
Lubbock,
TX 79404,
USA
Tel: 1 806 740 1600
Fax: 1 870 673 6319

PGS
See **Bayer** CropScience

Pioneer Hi-Bred International Inc.
7100 NW 62nd Avenue,
PO Box 1150,
Johnston,
IA 50131,
USA
Tel: 1 515 334 6908
Fax: 1 515 334 6886

Plant Genetic Systems
Josef Plateaustraat 22,
B-9000 Ghent,
BELGIUM
Tel: 32 9 235 8411
Fax: 32 9 224 0694
See **Bayer** CropScience

Rhône-Poulenc Secteur Agro
See **Bayer** CropScience

Rohm & Haas
See **Dow AgroSciences**

Rupp Seeds Inc.
17919 County Road B,
Wauseon,
OH 43567,
USA
Tel: 1 419 337 1841

Sandoz Agro Ltd
See **Syngenta**

Sands of Iowa
PO Box 468,
Marcus,
IA 51035-0648,
USA
Tel: 1 712 376 4135
Fax: 1 712 376 4140

Scotts Quality Seeds
1701 North Broadway,
Mt Pleasant,
IA 52641-0110,
USA
Tel: 1 319 385 8518

Seminis Vegetable Seeds Inc.
800 North Lindbergh Blvd,
Saint Louis,
MO 63167
USA
Tel: 1 314 694 1000
e-mail: vegetables@monsanto.com

Siebens Hybrids Inc.
633 North College Avenue,
Geneseo,
IL 61254,
USA
Tel: 1 309 944 5131
Fax: 1 309 944 6090

Societé National d'Exploitation des Tabacs et Allumettes
143 Blvd. Romain Rolland,
Paris,
F-75685 Cedex 14,
FRANCE
Tel: 33 01 44 97 61 50
Fax: 33 01 44 97 67 53
Home page: www.imperial-tobacco.com

Stine Seed
2225 Laredo Trail,
Adel,
IA 50003,
USA
Fax: 1 515 677 2716

Stoneville Pedigreed Seed Co.
6625 Lennox Park Drive,
Suite 117,
Memphis,
TN 38115,
USA
Tel: 1 901 375 5800
Fax: 1 901 375 5860

Syngenta
CH-4002,
Basel,
SWITZERLAND
Tel: 41 61 323 1111
Fax: 41 61 323 1212
Home page: www.syngenta.com/

Syngenta Crop Protection Inc.
PO Box 18300,
Greensboro,
NC 27419,
USA

Trisler Seed Farms Inc.
3274 East 800 North Road,
Fairmont,
IL 61841-6139,
USA
Tel/Fax: 1 217 288 9301

University of Hawaii,
3190 Maile Way,
Honolulu
HI 96822
USA

Vector Tobacco
1 Park Dr.
Ste. 150,
Research Triangle Park,
NC 27709,
USA
Tel: 1 212 687 8080,
Home page: www.vectorgroupltd.com

Wilson Seeds Inc.
PO Box 391,
Harlan,
IA 51537,
USA
Tel: 1 712 755 3841
Fax: 1 712 755 5261

Wrightson Seeds Ltd
Wrightson Research,
Kimihia Research Centre,
Tancreds Road,
Lincoln,
PO Box 939,
Christchurch,
NEW ZEALAND
Tel: 64 3 325 3158
Fax: 64 3 325 2417

Zeneca Agrochemicals
See **Syngenta**

Index

Approved names, common names, code numbers, Latin names and tradenames

This index lists alphabetically all approved names, common names, code numbers, Latin names and tradenames.

Latin names appear in italics (e.g. *Bacillus thuringiensis*) and tradenames appear in single quotation marks (e.g. 'NewLeaf').

All are referred to by entry number.